The Population Explosion
and Other Mathematical Puzzles

The
Population
Explosion
and Other
Mathematical
Puzzles

Dick Hess

 **World Scientific**

NEW JERSEY · LONDON · SINGAPORE · BEIJING · SHANGHAI · HONG KONG · TAIPEI · CHENNAI · TOKYO

Published by

World Scientific Publishing Co. Pte. Ltd.

5 Toh Tuck Link, Singapore 596224

USA office: 27 Warren Street, Suite 401-402, Hackensack, NJ 07601

UK office: 57 Shelton Street, Covent Garden, London WC2H 9HE

Library of Congress Cataloging-in-Publication Data
Names: Hess, Dick.
Title: The population explosion and other mathematical puzzles / Dick Hess.
Description: New Jersey : World Scientific, 2016.
Identifiers: LCCN 2015044541| ISBN 9789814740975 (hrd : alk. paper) |
    ISBN 9789814733755 (pbk. : alk. paper)
Subjects: LCSH: Mathematical recreations. | Mathematics--Problems, exercises, etc.
Classification: LCC QA95 .H4785 2016 | DDC 793.74--dc23
LC record available at http://lccn.loc.gov/2015044541

British Library Cataloguing-in-Publication Data
A catalogue record for this book is available from the British Library.

Typeset by Stallion Press
Email: enquiries@stallionpress.com

Printed in Singapore

To the loving memory of my dear brother,
Robert A. (Bob) Hess,
9 November 1940 to 2 March 2015.

Preface

This book is a sequel to *Mental Gymnastics: Recreational Mathematical Puzzles* and *Golf on the Moon*, written by me and published by Dover Publishing Co. in 2011 and 2014 respectively. The puzzles in all volumes are for the reader's enjoyment and should be passed on to others for their enjoyment as well. They are meant to challenge mathematical thinking processes, including logical thought, insight, geometrical, analytical and physical concepts, and may require considerable perseverance. While most of the puzzles can be solved by pencil and paper analysis, there are some that are best tackled with a computer to search for or calculate a solution. Be prepared to keep your wits about you.

I often encounter the ideas for many of these puzzles in publications or on-line sources that offer problem columns or puzzle sections. These include *Crux Mathematicorum with Mathematical Mayhem, Journal of Recreational Mathematics, Pi Mu Epsilon Journal, Puzzle Corner* in *Technology Review, Ponder This* and *Puzzle Up*. Other puzzle ideas were introduced to me by word of mouth through a delightful community of puzzle solvers. I owe a debt of gratitude to all enthusiasts who love to share their latest challenges and listen to mine.

Dick Hess

Contents

Chapter 1
Playful Puzzles

1
Word Mystery

What word has 8 letters, sometimes has 9 — it always contains 8 letters, occasionally uses 12 though! Find either of two answers.

2
Salary Secrecy

A group of 5 employees is at lunch and the subject of their average salary comes up. They all want to know the average but don't want to give information to any other about their own salary. Each has a pencil and piece of paper and there is no one else to assist them. How can they meet their objective?

3
Relations Puzzles

(a) A man points to another man and says: "Sons and daughters have I none but that man's father is my father's son." How are the two men related?

(b) Ray's son-in-law is my Uncle Bob's father. If I am related by blood to Ray how is Ray related to me?

(c) "Daughters and nephews have I none but Chris's father-in-law is my mother-in-law's son."

    (i) What is the speaker's gender?
    (ii) What is Chris's gender?
    (iii) How are the speaker and Chris related?

4
Slider

Use only five sliding block moves to get the piece labeled T to the lower right corner. A move is one piece moved along any path.

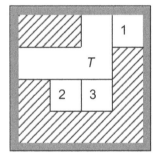

5
Fastest Serve

A tennis player hits a serve that in kilometers per hour (kph) is exactly 100 more than when expressed in miles per hour (mph). How fast did he serve?

6
The Population Explosion

In March 2015 the estimated population of the earth reached 7.3 billion people. The average person is estimated to occupy a volume of 0.063 m^3 so the volume of the total population is 0.4599 km^3.

(a) Model the earth as a sphere with a radius of 6,371 km and spread the volume of people over the surface of the earth in a shell of constant thickness. How thick is the shell?

(b) The population currently grows geometrically at 1.14% a year. How long will it take at this rate for the population to fill a shell one meter thick covering the earth? What will the population be then?

(c) At the 1.14% geometric rate how long will it take and what will the population be to occupy a sphere with a radius expanding at the speed of light $(= 9.4605284 \times 10^{12}$ km/yr)? Ignore relativistic effects.

7
Catenary

A 15 meter chain hangs from two vertical 10 meter poles placed d meters apart. The low point of the chain hangs 2.5 meters from the ground. What is d?

Chapter 2

Geometric Puzzles

8
Mining on Rigel IV

An amazing thing about the planet Rigel IV is that it is a perfectly smooth sphere of radius 4,000 miles. Like the earth it rotates about a north pole so a latitude and longitude system of coordinates referenced to the poles serves to locate positions on Rigel IV just as it does on earth. Three prospectors make the following reports to headquarters.

(a) Prospector A: "From my base camp I faced north and went 1 mile in that direction without turning. Then I went east for 1 mile. I rested for lunch before facing north again and going 1 mile in that direction without turning. Finally, I went west for 1 mile and arrived exactly at my base camp." What are the possible locations for base camp A?

(b) Prospector B: "From my base camp I went 1 mile north; then I went 1 mile east. I next went 1 mile south and, finally, I went 1 mile west and arrived exactly at my base camp." What are the possible locations for base camp B?

(c) Prospector C: "From my base camp I went 1 mile north; then I went 1 mile east. I next went 1 mile south and, finally, I went 1 mile west and arrived at the most distant point possible from my base camp under these conditions." What are the possible locations for base camp C and how far from base camp C does the prospector end up?

9
Linking Points

A straight line connecting any two points of the six points in the figure is a link. How many links can be placed without forming a triangle with three of the points as vertices?

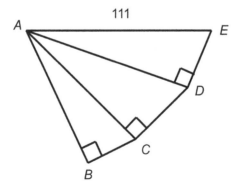

10
Right Triangles

In the figure $AE = 111$ and other lengths are unknown. What is the value of $AB^2 + BC^2 + CD^2 + DE^2$?

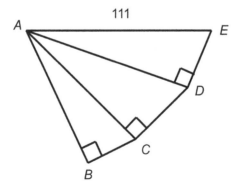

11
The Clipped Polyhedron

A polyhedron, P_1, has a small tip of each of its vertices sliced off by a plane to produce polyhedron P_2. P_2 has F faces, V vertices and E edges.

(a) One of F, V or E equals 11. What are the two possibilities for P_1?

(b) One of F, V or E equals 13. What are the four possibilities for P_1?

12
The Papered Boxes

(a) Aksana states to her friend, Josh: "I have an ideal rectangular box with integer dimensions and no top. I have papered both inside and outside (ten surfaces) and notice that, amazingly, the area of paper in square units is the same as the volume of the box in cubic units. Furthermore, my box has the maximum volume for this situation." Josh replies: "My box has the same properties but the minimum volume possible." What are the dimensions of the two boxes?

(b) Bob states to Kathy: "I have an ideal rectangular box with integer dimensions and no top. I have papered the five outside surfaces and notice that, amazingly, the area of paper in square units is the same as the volume of the box in cubic units. Furthermore, my box has the volume of a cube but is not a cube." Kathy replies: "My box also has an area of paper in square units that is the same as the volume of the box in cubic units but has half the volume of yours." What are the dimensions of the two boxes?

(c) Chris states to her friend, Laurie: "I have an ideal rectangular box with integer dimensions. I have papered both inside and outside of the six surfaces and notice that, amazingly, the area of paper in square units is the same as the volume of the box in cubic units. Furthermore, the largest dimension of my box is odd." Laurie replies: "My box has the same properties but a smaller volume than yours." What are the dimensions of the two boxes?

(d) David states to his friend, Mary: "I have an ideal rectangular box with integer dimensions. I have papered the six outside surfaces and notice that, amazingly, the area of paper in square units is the same as the volume of the box in cubic units. Furthermore, the dimensions of my box are all different." Mary replies: "My box has the same properties but half the volume of yours." What are the dimensions of the two boxes?

13
The Almost Rectangular Lake

The figure shows a lake *ABCDEFG* that is nearly rectangular except for the segment of its shoreline, *DE*. Let A_1 be the area of *DEFG*, let A_2 be the area of *BCDE* and let A_3 be the area of the wedge *ADE*. What is A_3 in terms of A_1 and A_2?

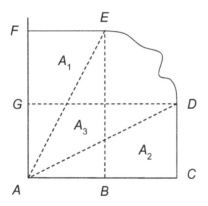

14
The Trifurcated Diamond

The tile shown is a diamond with 120° and 60° angles. Divide it into three tiles that are similar to each other, that is, they are the same shape but may be of different sizes. Tiles may be turned over. Find six solutions.

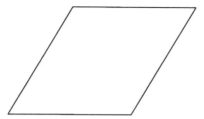

15
Square Dissection

Completely dissect a square into the lowest number of different sized rectangles with integer edges and a length to width ratio of 3 to 1.

Chapter 3
Digital Puzzles

16
Pandigital Purchases

Last Christmas I purchased several gifts, each costing a perfect square number of dollars. When I write down the set of prices every integer from 1 to 9 is represented exactly once. If my total cost was the minimum possible, what was the total bill and how many gifts did I buy?

17
Sudoku Variant

Place the numbers 1 to 5 in the grid so that no number is repeated in any row or column and the sums of numbers in the outlined regions are all different.

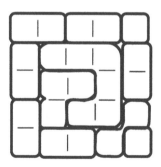

18
Pandigital Sums

(a) Notice that in the expression $87/93 + 42/651 = 1$ the left-hand side uses each digit from 1 to 9 exactly once. Find positive integers a to d in the expression $a/b + c/d = 1$ with this property where:

   (i) $a+b+c+d$ is maximized;
   (ii) $a+b+c+d$ is minimized;
   (iii) a/b is minimized.

(b) In the expression $70/96 + 143/528 = 1$ the left-hand side uses each digit from 0 to 9 exactly once. Find positive integers a to d in the expression $a/b + c/d = 1$ with this property where:

   (i) $a+b+c+d$ is maximized;
   (ii) $a+b+c+d$ is minimized;
   (iii) a/b is minimized.

(c) In the expression $4/68 + 297/153 = 2$ the left-hand side uses each digit from 1 to 9 exactly once. Find positive integers a to d in the expression $a/b + c/d = 2$ with this property where:

   (i) $a+b+c+d$ is maximized;
   (ii) $a+b+c+d$ is minimized;
   (iii) a/b is minimized.

(d) In the expression $12/43 + 870/465 = 2$ the left-hand side uses each digit from 0 to 9 exactly once. Find positive integers a to d in the expression $a/b + c/d = 2$ with this property where:

   (i) $a+b+c+d$ is maximized;
   (ii) $a+b+c+d$ is minimized;
   (iii) a/b is minimized.

19
Even and Odd

Let m be a 5-digit number containing each of the odd digits 1, 3, 5, 7 and 9 in some order and let n be a 5-digit number containing each of the even digits 2, 4, 6, 8 and 0 in some order. Is it possible that n is a multiple of m?

20
Integer Oddity

I have two different single digits, a and b. If I take the integers closest to the expressions $^{-0.a}\!\!\sqrt{0.b}$ and $^{-0.b}\!\!\sqrt{0.a}$, I notice that I get two different integers, each greater than 10, where the digits are permutations of each other. What are a and b?

21
A 10-Digit Number

I have a 10-digit number made up of the digits 0, 1, 2, 3, 4, 5, 6, 7, 8 and 9 in some permutation. When digits are removed one at a time from the left, the number remaining is divisible in turn by 9, 8, 7, 6, 5, 4, 3, 2 and 1.

(a) My number is the largest possible for this to happen; what is it?

(b) My number is the smallest possible for this to happen; what is it?

Chapter 4

Logical Puzzles

22
Cake Division

(a) Joe and Bob divide a cake. Joe cuts the cake into two pieces; then Bob cuts one of those pieces into two pieces. Joe gets the largest and smallest of the 3 pieces produced. What is the largest fraction of cake Joe can guarantee himself and how large is it?

(b) Now suppose that both Joe and Bob are on a diet and wish to receive as little cake as possible. How much cake can Bob force Joe to take?

(c) Now suppose Joe cuts the cake into two pieces; then Bob cuts one of those pieces into two pieces. Then Joe cuts one of the three pieces into two pieces. Joe gets the largest and smallest pieces produced; Bob gets the middle two pieces. What is the largest fraction of cake Joe can guarantee himself and how large is it?

23
Prisoner's Escape

The knight is told by his evil jailer: "Each square of the 7 by 17 grid contains a hidden number. Tell me the sum of all 119 numbers and you go free; otherwise you will be executed. You will get to uncover one number and have one hour to prepare your strategy." Upon being taken back to his cell the knight is told by the guard that each 3 by 4 or 4 by 3 rectangle anywhere in the grid has numbers adding up to 202. How will the knight escape?

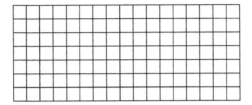

24
Logical Question

In the figure, showing a square and an equilateral triangle, is *s* larger or smaller than the radius of the circumscribed circle?

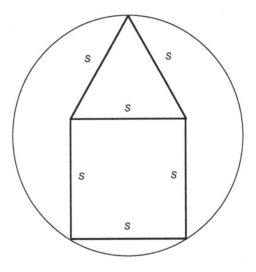

25
Ali Baba and the 10 Thieves

Ten thieves, ranked *A* to *J*, are trying to cross a river in a boat requiring two rowers. Unfortunately, if the ranks of any two in the boat differ by more than 1, those two will refuse to stay in the boat. This constraint means they can't get across the river. Their leader, with a rank of *A*, asks Ali Baba for help and Ali Baba replies, "If you give me a rank of *A*, equal to yours, we can all cross the river." The leader agrees. How many one-way crossings are the least required to get Ali Baba and the 10 thieves across the river?

26
Anniversary Party Puzzles

This set of puzzles imagines an anniversary party where a couple makes statements about their children's ages. The children are assumed to be born in wedlock so that none is as old as the anniversary being celebrated. In each problem the couple states: "Earlier tonight we gave Smith the sum and product of the ages of our 3 children but Smith didn't get their ages right." The couple then continues with various follow up statements for each of the puzzles given below. Each party is given by a different couple so there is no connection between one puzzle and another. Smith, Jones and Brown are assumed to be error-free in their logic. In each puzzle determine how old the children are.

(a) At a 9th anniversary party couple *A* also states: "Jones missed the problem 1 year ago."

(b) At a 30th anniversary party couple *B* also states: "Jones missed the problem 1 year ago and Brown missed it 3 years ago."

(c) At a 30th anniversary party couple *C* also states: "Jones missed the problem after hearing Smith's answer tonight and Brown missed it 4 years ago."

(d) At a 30th anniversary party couple *D* also states: "Jones missed the problem 4 years ago and Brown missed it 7 years ago."

(e) At a 30th anniversary party couple *E* also states: "Jones missed the problem 5 years ago and Brown missed it 6 years ago."

(f) At a 30th anniversary party couple *F* also states: "Jones missed the problem 2 years ago and Brown missed it 10 years ago."

(g) At a 28th anniversary party couple *G* also states: "Jones missed the problem 2 years ago and Brown missed it 6 years ago."

(h) At a 28th anniversary party couple *H* also states: "Jones missed the problem 4 years ago and Brown missed it 8 years ago."

The next set of puzzles has the same conditions except the couple gives Smith and Jones the sum of the ages and the sum of the cubes of the ages of the three children on the night of the party.

(i) At a 25th anniversary party couple *I* also states: "Jones missed the problem 1 year ago."

(j) At a 20th anniversary party couple *J* also states: "Jones missed the problem 2 years ago."

(k) At a 25th anniversary party couple *K* also states: "Jones missed the problem 3 years ago."

(l) At a 25th anniversary party couple *L* also states: "Jones missed the problem 4 years ago."

(m) At a 30th anniversary party couple *M* also states: "Jones missed the problem 9 years ago."

Chapter 5

Probability Puzzles

27
Gambler's Surprise

You have 100 cards; 75 are marked "win" and 25 are marked "lose". You start with $10,000 and must bet 90% of your remaining money at 1 to 1 odds on each of the 100 cards in turn. At the end how much money will you have left? Suppose instead there were 80 "win" cards and 20 "lose" cards then how much money would you have left?

28
Tenzi

In the game of Tenzi 10 ordinary dice are rolled and a target number is determined as the number most frequently appearing in that roll. On subsequent rolls you save out all dice with the target number showing and roll the remainder. You stop when all 10 dice show the target number. What is the expected number of rolls in Tenzi before stopping?

29
Little Bingo

Consider the game of Little Bingo, where the cards have only three columns and rows as shown with a free center cell. Columns are labeled B, N and G. Cells in the B column contain unrepeated numbers from 1 to 6; cells in the N column contain unrepeated numbers from 7 to 12 and cells in the G column contain unrepeated numbers from 13 to 18. There are $T_6 = 6^3 \times 5^3 \times 4^2 = 1,728,000$ such cards possible. Imagine a Bingo hall with 1,728,000 players, each having a different card. The game is played by drawing at random from among the 18 numbers without replacing them. A card is a winner if it has a completed row, column or diagonal from the numbers drawn so far. The player holding that card yells "Bingo" as soon as that card is a winner and the game stops.

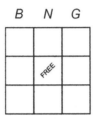

(a) How many draws are required before a winner is guaranteed?

(b) There will always be multiple winners when the game ends and the group of winners will either all win on a row or a diagonal, or they will all win on a column. Call these possibilities a "Row Win" and a "Column Win". What is the probability of a "Row Win" in Little Bingo?

30
Regular Bingo

In the game of Regular Bingo the cards have five columns and rows as shown with a free center cell. Columns are labeled B, I, N, G and O. Cells in the B column contain unrepeated numbers from 1 to 15; cells in the I column contain unrepeated numbers from 16 to 30 and so on until cells in the O column contain unrepeated numbers from 61 to 75. There are $T_{15} = (15 \times 14 \times 13 \times 12)^5 \times 11^4 = 5.52446 \times 10^{26}$ such cards possible. Imagine a Bingo hall with T_{15} players, each having a different card. The game is played by drawing at random from among the 75 numbers without replacing them. A card is a winner if it has a completed row, column or diagonal from the numbers drawn so far. The player holding that card yells "Bingo" as soon as that card is a winner and the game stops.

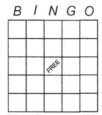

(a) How many draws are required before a winner is guaranteed?

(b) There will always be multiple winners when the game ends and the group of winners will either all win on a row or a diagonal, or they will all win on a column. Call these possibilities a "Row Win" and a "Column Win". What is the probability of a "Row Win" in Regular Bingo?

31
Fair Duel

Smith and Brown have challenged each other to a duel. They will take turns shooting at one another until one has been hit. Smith, who can hit Brown only 40% of the time, is the weaker shot so will be allowed to go first. They have determined that the duel favors neither shooter. What is Brown's probability of hitting Smith?

32
A Golden Set of Tennis

In tennis a golden set occurs when one player wins the first 24 points of a set. In "Grand Slam" tournaments women play the best 2 sets of 3 and men play the best 3 sets of 5. What is the probability of a golden set somewhere in the match if points are determined by a coin flip for (a) a women's match and for (b) a men's match?

33
Bus Ticket Roulette

(a) You are stuck in Las Vegas with only $2. Your only way out of town is to get enough money for the $4 bus ticket and your only way of getting the $4 is to place bets of whole dollar amounts on the roulette table. What is your best betting strategy and what is your probability of getting out of town?

(b) Your problem is the same as in part (a) except now the ticket out of town is $5.

Allowable bets are shown in the roulette diagram.

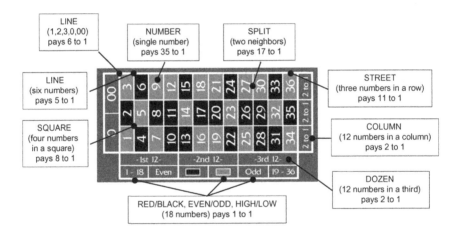

34
Boxes of Colored Balls

(a) A box contains a total of n balls. The balls are of two different colors and the numbers of balls of these colors are n_1 and n_2. An experiment is conducted in which balls are drawn from the box randomly without replacement until only one color remains. What is the expected number of balls in the box at the end of the experiment?

(b) Now the box contains balls of three different colors and the numbers of balls of these colors are n_1, n_2 and n_3. An experiment is conducted in which balls are drawn from the box randomly without replacement until only one color remains. This ends the experiment. At the outset of the experiment, the expected number of balls remaining in the box at the end of the experiment was calculated and found to be an integer. It was also noted that n_1, n_2 and n_3 are relatively prime in pairs. What are the smallest values possible for n_1, n_2 and n_3 and what is the expected number of balls in the box at the end of the experiment?

(c) Again the box has balls of three colors. In this problem the experiment ends when the last ball of any one of the colors has been drawn. This leaves the box containing balls of the remaining two colors. At the outset of the experiment, the expected number of balls remaining in the box at the end of the experiment was calculated and found to be an integer. How many balls of each color were initially in the box and what is the expected number of balls in the box at the end of the experiment?

35
Candyland

Imagine you have a board game in which you start on square 0 of a long track, as in the game of Candyland, that ends at square M. On each turn you roll a fair six-sided die and advance that many spaces. The game ends when you reach or exceed square M. You decide to play 21 such games. If M is larger than 100 what is your expected number of rolls to the nearest integer to complete the 21 games?

Chapter 6

Analytical Puzzles

36
Airport Rush

You are walking to your gate in the airport and notice you must stop for 1 minute to tie your shoe. You realize that you will soon come to a moving walkway. Should you wait to tie your shoe while on the walkway or does it make no difference in getting to your gate at the earliest possible time? Assume you walk at a constant speed relative to the ground or walkway when you are walking.

37
Which Meal?

Just as Kevin starts a meal he notices the minute hand is on a second mark exactly one second ahead of the hour hand. Which meal is Kevin enjoying?

38
Rat Race

Three cats and a rat are confined to the edges of a tetrahedron. The cats are blind but catch the rat if any cat meets the rat. One cat can travel 1% faster than the rat's top speed and the other two cats can travel 1% faster than half the rat's speed. Devise a strategy for the cats to catch the rat.

39
Family Visits

(a) Three brothers, Adam, Bill and Charles, plan a trip to visit Grandpa. They have a racing bicycle among them and 3 hours to get to Grandpa's. Adam is the only one allowed to pedal the bicycle and pedals at a speed of 40 mph when he has no passenger and 30 mph if he's carrying a passenger. He can carry no more than one passenger. On foot, Bill can go 13 mph and Charles can go 9 mph. How far away can Grandpa's be while still allowing the brothers to arrive at the 3-hour time limit?

(b) Two teams of three cousins each plan a trip to visit Aunt Jenny. Each team has a racing bicycle among them and 3 hours to get to Aunt Jenny's. Each has only one member allowed to pedal the bicycle who pedals at a speed of 25 mph when he has no passenger and 20 mph if he's carrying a passenger. The bicyclist can carry no more than one passenger. Each team has a runner who can go 10 mph on foot and a walker who can go 5 mph on foot. Team A starts by having the walker set out on foot first. Team B has the runner starting out on foot first. If Aunt Jenny lives 42 miles away and Team A arrives at Aunt Jenny's in exactly 3 hours, how late will Team B be to Aunt Jenny's?

40
Logging Problem

Define x as $x = \log_{16} 7 \times \log_{49} 625$. Find an expression for $\log_{10} 2$ in terms of x where the only constants appearing are integers.

41
Polynomial Problem 1

Define $P(x, y)$ as $P(x, y) = a_0 + a_1 x + a_2 y + a_3 x^2 + a_4 xy + a_5 y^2 + a_6 x^3 + a_7 x^2 y + a_8 xy^2 + a_9 y^3$.

Suppose $P(0, 0) = P(1, 0) = P(-1, 0) = P(0, 1) = P(0, -1) = P(1, 1) = P(1, -1) = P(12, 12) = 0$.

Find a point $(x_0, y_0) = (a/c, b/c)$ where a, b and $c > 1$ are integers and $P(x_0, y_0) = 0$ for all such polynomials defined above.

42
Polynomial Problem 2

The equation $xy + x + 5y + 13 = 8x^3$ has an integer solution of $(x, y) = (1, -1)$.

(a) How many other solutions are there where both x and y are integers?

(b) Find the solutions with both x and y primes. Count the negative of a prime as prime.

(c) Find the solutions with both x and y negative integers.

43
Cascaded Prime Triangles

The figure below (not drawn to scale) shows three Pythagorean triangles with a common side where, as shown in *All-Star Mathlete Puzzles* by Dick Hess, 2009, there is a solution with $x_0 = 271$, $x_1 = 36{,}721$, $x_2 = 674{,}215{,}921$.

(a) Find another solution with three cascaded prime Pythagorean triangles.

(b) Find an example with four Pythagorean triangles cascaded in the same fashion so that x_0 to x_4 are all prime numbers.

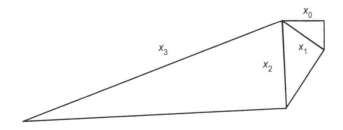

44
Egg Farming

Abel and Barry each manage egg farms.

Abel: "On my farm $1\frac{1}{7}$ hens lay $1\frac{1}{6}$ eggs in $1\frac{1}{5}$ days."

Barry: "On my farm $1\frac{1}{5}$ hens lay $1\frac{1}{6}$ eggs in $1\frac{1}{7}$ days."

(a) Which farm has hens that are more productive?

(b) Abel has 48 hens. How many days must he wait to first get a whole number of eggs exactly at the close of a day? How many eggs are produced by Abel's farm in that time?

(c) Barry has 300 hens. How many days must he wait to first get a whole number of eggs exactly at the close of a day? How many eggs are produced by Barry's farm in that time?

45
The Knight's Dilemma

"As a reward for your brilliance," said the king, "I give you the land you can walk around in a day. Take some of these stakes with you, pound them into the ground along your way, and be back at your starting point in 24 hours. All the land inside the polygon formed by the stakes will be yours." The knight takes n seconds to pound in each stake and notes that the regular polygonal path he takes to maximize his land has n sides. How much land does the knight get if he walks at a constant speed of 100 feet per minute?

46
The Least Equilateral Triangle

Pictured is an arbitrary triangle, *ABC*, with lines drawn to produce an internal triangle, $A'B'C'$. The angles (α, β, γ) are defined by $\alpha = A/n$, $\beta = B/n$ and $\gamma = C/n$ where n is an integer > 2. Note that if $n = 3$ then $A'B'C'$ will be equilateral for any choice of *ABC*. This is known as Morley's Theorem.

Your challenge is to pick *A*, *B*, *C* and *n* so that *A'B'C'* deviates from being equilateral by maximizing the expression $f = |A'-B'| + |A'-C'| + |B'-C'|$, where *A'*, *B'* and *C'* are the angles of *A'B'C'*. Find the least upper bound for *f* and determine the angles for the limiting triangle *A'B'C'* corresponding to that upper bound.

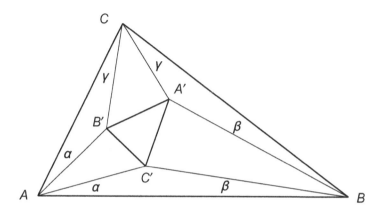

47
You're the Doctor

(a) You have an empty 5 ml vial, an empty 4 ml vial, a water supply, a sink and a single water-soluble tablet of medicine that you can add to a vial any time you choose. For which positive integers $n < 100$ is it possible to measure out a dose (dissolved in water) of exactly *n*% of a tablet.

Answer the same question for

(b) Vials of capacities 5 ml and 1 ml.

(c) Vials of capacities 9 ml and 1 ml.

(d) Vials of capacities 8 ml and 5 ml.

(e) Vials of capacities 12 ml and 5 ml.

Chapter 7

Physics Puzzles

48
Boating Surprise

A man is in a boat floating in a swimming pool. A large rock in the boat is tossed overboard into the water. The water level of the swimming pool on its sides settles to the same place it started. Explain how.

49
Balance Problems

Each puzzle shows an assembly of weights hanging from a peg on the wall. Pick whole numbered weights without repeat from the designated collection so that all elements balance. The beams are assumed to be weightless.

(a) In the figure balance it all by finding unique weights A to E from among the values 1–115.

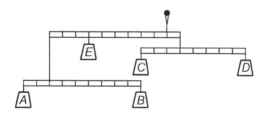

(b) In the figure balance it all by finding unique weights *A* to *F* from among the values 1–75.

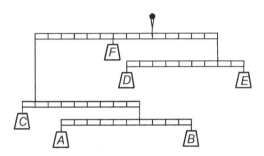

(c) In the figure balance it all by finding unique weights *A* to *F* from among the values 1–794.

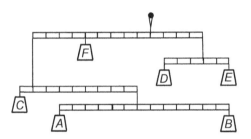

(d) In the figure balance it all by finding unique weights *A* to *F* from among the values 1–140.

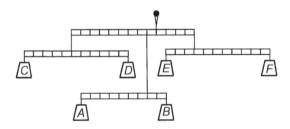

(e) In the figure balance it all by finding unique weights *A* to *F* from among the values 1–140.

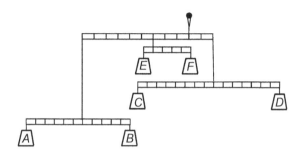

(f) In the figure balance it all by finding unique weights *A* to *F* from among the values 1–135.

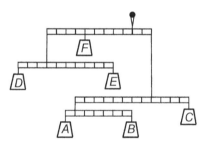

(g) In the figure balance it all by finding unique weights *A* to *F* from among the values 1–190.

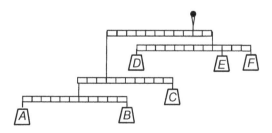

(h) In the figure balance it all by finding unique weights *A* to *F* from among
     the values 1–65.

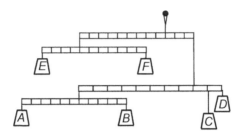

(i) In the figure balance it all by finding unique weights *A* to *F* from among
     the values 1–130.

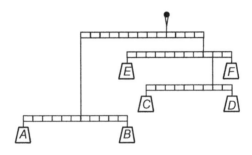

(j) In the figure balance it all by finding unique weights *A* to *E* from among
     the values 1–30.

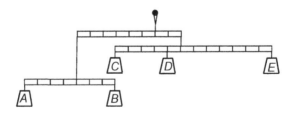

Each puzzle in this next group shows an assembly of weights hanging from a peg on the wall. Each beam weighs one unit per block with weight distributed uniformly along the beam.

(k) In the figure balance it all by finding unique weights A to F from among the values 1–92.

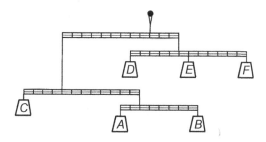

(l) In the figure balance it all by finding unique weights A to F from among the values 1–800.

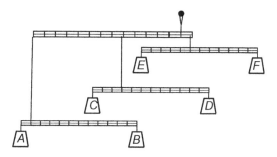

(m) In the figure balance it all by finding unique weights A to F from among the values 1–270.

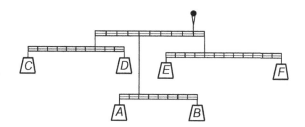

(n) In the figure balance it all by finding unique weights *A* to *F* from among the values 1–70.

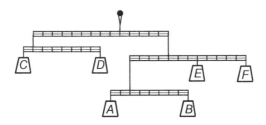

50
The Hanging Rod

A uniform thin rod, *BC*, of length 3 units is hanging from a ceiling by supports at *A* and *D*, 5 units apart. The supporting strings, *AB* and *CD* are of lengths 2 and 4 as shown.

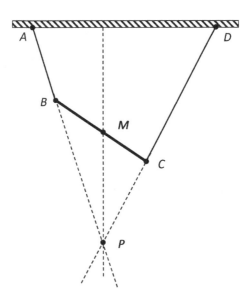

(a) Show that the downward projection of the center of the rod, *M*, intersects the extension of the two strings at *P*.

(b) Find the internal angles *A*, *B*, *C* and *D*.

51
Falling Ladders

A uniform thin ladder starts as shown propped up at angle θ_0 between two frictionless walls separated by a frictionless floor. It is slotted in the floor and wall at B and wall at U to ensure the ends don't leave those surfaces. The ladder does not bend and is in a uniform gravity field of $g = 32$ ft/sec^2. A test mass is suspended from the ceiling at M, which is at the same height as U. Students are allowed to move the walls to adjust θ_0 and the height of the mass so that it is always at the same height as the top of the ladder. Three students report on experiments they performed.

Student A: "I simultaneously removed both walls and cut the string supporting the mass. The mass and ladder hit the floor at the same instant."

Student B: "I simultaneously removed the right wall and cut the string supporting the mass. The mass and ladder hit the floor at the same instant."

Student C: "I simultaneously removed the left wall and cut the string supporting the mass. The mass and ladder hit the floor at the same instant."

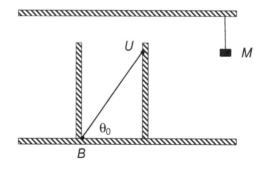

(a) What value of θ_0 did student A use?

(b) What was the difference in values of θ_0 used by students B and C?

Chapter 8

Trapezoid Puzzles

Integer-Sided Trapezoids Inscribed In Circles

The next several puzzles relate to integer-sided isosceles trapezoids (IIT's) that can be inscribed in a circle with integer radius. There are two types as shown in the figure. An isosceles trapezoid is a quadrilateral with parallel sides, $a < b$, and equal slant sides, c. A computer search will be helpful for some of these puzzles.

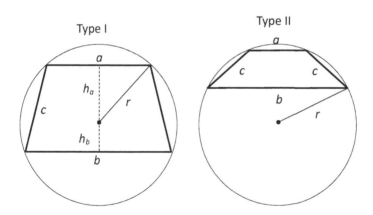

52
Smallest IIT's

(a) Find the IIT that can be inscribed in a circle with smallest radius.

(b) What is the IIT with the next smallest r which has $a = 1$?

(c) What is the IIT with next smallest r and with $a = 1$ and $b = 2r$?

(d) Find an infinite set of IIT's with $a = 1$ and $b = 2r$.

53
Smallest Circle

(a) Find the smallest circle with integer radius that can have an IIT with $a = 1$ and both b and c even numbers.

(b) Find the case with the next smallest r.

54
Prime IIT's

(a) Find two IIT's with r, b, a and c all having different prime numbers.

(b) * Is there another one? (* a solution is not known).

(c) Find IIT's with r, b, a and c all having prime numbers, possibly the same.

(d) * Are there any with $r \neq c$? (* a solution is not known).

55
Flat IIT's

(a) Find an IIT with $c = 1$ and $r > 10$.
(b) Find an infinite series of IIT's with $c = 1$.

56
Pointed IIT's

(a) Find the smallest r for an IIT for which $c/b > 5$.

(b) Find the smallest r for an IIT for which $c/b > 10$.

(c) Find an infinite set of IIT's with c/b increasing arbitrarily.

57
Nearly Square IIT's

Consider IIT's that are nearly square and use $q = (b - a + |c - b|)/b$ as a measure of how close the IIT is to a square. (The smaller q is the closer the IIT is to a square.)

(a) What are the two smallest IIT's with $q < 0.2$?

(b) Find a series of IIT's with q tending to 0 as the series continues.

58
IIT's with Integer Altitudes

Find the IIT with the smallest r such that the altitudes h_a and h_b are both integers.

Integer-Sided Trapezoids Inscribed in Squares

The next several puzzles relate to integer-sided isosceles trapezoids (IIT's) that can be inscribed in a square with integer side, s. There are four types as shown in the figure. An isosceles trapezoid is a quadrilateral with parallel sides, $a < b$, and equal slant sides, c. A computer search will be helpful for some of these puzzles.

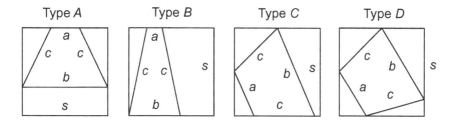

59
Smallest IIT's in a Square

(a) Find the IIT that can be inscribed in a square with the smallest side.

(b) Find the IIT that can be inscribed in a square with the smallest side for solution Types B, C and D.

(c) Find the IIT with $c > b$ that can be inscribed in a square with the smallest side for solution Type A.

(d) Find the Type A IIT with $b/c \leq 0.9$ that can be inscribed in a square with the smallest side.

60
IIT's with $a = s$

(a) Find the IIT with $a = s$ that can be inscribed in a square with the smallest side.

(b) Find an infinite series of IIT's with $a = s$ that can be inscribed in squares.

61
IIT's with a=d

(a) Find the IIT (not Type A) with $a = d$ that can be inscribed in a square with the smallest side.

(b) Find an infinite series of IIT's (not Type A) with $a = d$ that can be inscribed in squares.

62
IIT's with $x = u$

(a) Find the Type C IIT with $x = u$ that can be inscribed in a square with the smallest side.

(b) Find an infinite series of Type C IIT's with $x = u$ that can be inscribed in squares.

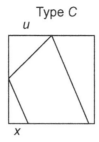

Type C

u

x

63
IIT's with *b/s* > 1.4

(a) Find the Type *C* IIT with *b/s* > 1.4 that can be inscribed in a square with the smallest integer side.

(b) Find an infinite series of Type *C* IIT's with *b/s* getting successively closer to $\sqrt{2}$ that can be inscribed in integer-sided squares.

(c) Find the Type *D* IIT with *b/s* > 1.4 that can be inscribed in a square with the smallest integer side.

(d) Find an infinite series of Type *D* IIT's with *b/s* getting successively closer to $\sqrt{2}$ that can be inscribed in integer-sided squares.

64
IIT's with *b/s* < 0.8, 0.75 and 0.71

(a) Find the IIT with *b/s* < 0.8 that can be inscribed in a square with the smallest integer side.

(b) Find the IIT with *b/s* < 0.75 that can be inscribed in a square with the smallest integer side.

(c) Find the IIT with *b/s* < 0.71 that can be inscribed in a square with the smallest integer side.

65
IIT's with Minimum Coverage

(a) Find the smallest square with integer side containing an IIT with an area less than 0.5000005 of the square.

(b) Find an infinite series of such cases with Area/s^2 getting successively closer to 1/2.

66
Largest Square

Find the largest square with an integer side that cannot circumscribe a Type D IIT.

67
Multiple Solutions

Find the smallest IIT that is inscribable in more than one integer-sided square.

68
Smallest Squares

(a) Find the smallest integer-sided square that allows more than one IIT to be inscribed in it.

(b) Find the smallest integer-sided square that allows 45 IIT's to be inscribed in it.

(c) Find the smallest integer-sided square with an IIT inscribed in it such that $a \leq c \leq b$ and $(b - a)/a \leq 0.01$.

(d) Find the smallest integer-sided square with a Type D IIT inscribed in it such that $a \leq c \leq b$ and $(b - a)/a \leq 0.01$.

(e) Find an infinite sequence of integer-sided squares with Type D IIT's inscribed in them such that $(b - a)/a$ gets successively smaller and approaches 0 as the sequence proceeds.

(f) What is the smallest integer-sided square that allows more than one IIT of Types C or D to be inscribed within it?

(g) Find the smallest integer-sided square that allows a Type D IIT to be inscribed in it such that $b = s$.

(h) Find the smallest integer-sided square that allows two different Type D IIT's to be inscribed in it such that each has $b = s$.

(i) Find the two smallest squares with odd integer sides that can circumscribe an IIT that is not Type A. The IIT's are different for the two cases.

Chapter 9

Jeeps in the Desert

The problems in this chapter deal with fleets of jeeps in the desert initially located at point A where there is a fuel depot with unlimited fuel supply. All jeeps end up at either point A or at a delivery point B, as far from point A as possible. The jeeps are all identical, can go a unit distance on a tank of fuel and consume fuel at a constant rate per mile. Jeeps may not tow each other or carry more than a tankful of fuel.

69
One-Way Trip with a Single Jeep

Your fleet consists of one jeep, which is trying to get to point B as far away as possible from the depot and finish at point B. It is permitted to cache fuel unattended in the desert for later use.

(a) How far can you get if you may use only 2 tanks of fuel?

(b) How far can you get if you may use only 1.9 tanks of fuel?

(c) How much fuel is required to go 1.33 units of distance?

(d) How far can you get if you may use only 3 tanks of fuel?

(e) How far can you get if you may use only 2.5 tanks of fuel?

(f) How much fuel is required to go 1.5 units of distance?

(g) How much fuel is required to go 2 units of distance?

70
Round Trip with a Single Jeep

Your fleet consists of one jeep, which must deliver a package to point B as far away as possible from the depot and then return to the depot. It is permitted to cache fuel unattended in the desert for later use.

(a) If you may use only 3 tanks of fuel what is the maximum distance of point B?

(b) If you may use only 6 tanks of fuel what is the maximum distance of point B?

(c) How much fuel is required if point B is 0.7 units away?

(d) If you may use only 2.5 tanks of fuel what is the maximum distance of point B?

(e) How much fuel is required if point B is 1 unit away?

(f) How much fuel is required if point B is 1.5 units away?

71
A Single One-Way Trip with Two Jeeps

You have a fleet of two jeeps. Jeep 1 must deliver a package to point B as far away as possible from the depot. Jeep 1 finishes at B; Jeep 2 provides support and finishes at A. Fuel may be transferred only between the two jeeps; no caching is permitted.

(a) If you may use only 2 tanks of fuel what is the maximum distance of point B?

(b) If you may use only 2.5 tanks of fuel what is the maximum distance of point B?

(c) How much fuel is required if point B is 13/9 units away?

(d) How much fuel is required if point B is 1.47 units away?

(e) How much fuel is required if point B is 1.495 units away?

72
Two One-Way Trips with Two Jeeps

You have a fleet of two jeeps. Each must deliver a package to point B as far away as possible from the depot. Both jeeps finish at B. Fuel may be transferred only between the two jeeps; no caching is permitted.

(a) If you may use only 3 tanks of fuel what is the maximum distance of point B?

(b) How much fuel is required if point B is 11/9 units away?

(c) How much fuel is required if point B is 1.23 units away?

(d) If you may use only 5 tanks of fuel what is the maximum distance of point B?

(e) If you may use only 4.95 units of fuel what is the maximum distance of point B?

(f) How much fuel is required if point B is 1.2475 units away?

73
One Round Trip with Two Jeeps

You have a fleet of two jeeps. Jeep 1 must deliver a package to point B as far away as possible from the depot and return to the depot. Jeep 2 provides support and finishes at the depot. Fuel may be transferred only between the two jeeps; no caching is permitted.

(a) If you may use only 3 tanks of fuel what is the maximum distance of point B?

(b) If you may use only 3.5 tanks of fuel what is the maximum distance of point B?

(c) How much fuel is required if point B is 0.97 units away?

(d) How much fuel is required if point B is 53/54 units away?

(e) How much fuel is required if point B is 59/60 units away?

(f) If you may use only 7.8 tanks of fuel what is the maximum distance of point B?

(g) How much fuel is required if point B is 99/100 units away?

74
Two Round Trips with Two Jeeps

You have a fleet of two jeeps. Both jeeps must deliver a package to point B as far away as possible from the depot and return to the depot. Fuel may be transferred only between the two jeeps; no caching is permitted.

(a) If you may use only 4 tanks of fuel what is the maximum distance of point B?

(b) How much fuel is required if point B is 0.7 units away?

(c) If you may use only 6 tanks of fuel what is the maximum distance of point B?

(d) If you may use only 7 tanks of fuel what is the maximum distance of point B?

(e) How much fuel is required if point B is 0.74 units away?

(f) If you may use only 9.2 tanks of fuel what is the maximum distance of point B?

(g) If you may use only 9.8 tanks of fuel what is the maximum distance of point B?

(h) How much fuel is required if point B is 121/162 units away?

75
Two Jeeps and Two Depots

You have a fleet of two jeeps. Both jeeps must deliver a package to point B as far away as possible from point A. There are fuel depots at points A and B. Fuel may be transferred only between the two jeeps; no caching is permitted.

(a) How much fuel is required if point *B* is 1.25 units away?

(b) How much fuel is required if point *B* is 1.34 units away?

(c) If you may use only 5.02 tanks of fuel what is the maximum distance of point *B*?

(d) If you may use only 6.9 tanks of fuel what is the maximum distance of point *B*?

(e) How much fuel is required if point *B* is 1.475 units away?

(f) If you may use only 8.6 tanks of fuel what is the maximum distance of point *B*?

(g) If you may use only 9.992 tanks of fuel what is the maximum distance of point *B*?

76
A Single One-Way Trip with Three Jeeps

You have a fleet of three jeeps. Jeep 1 must deliver a package to point *B* as far away as possible from a single depot at point *A* and finish at point *B*. Jeeps 2 and 3 provide support and finish at point *A*. Fuel may be transferred only between the jeeps; no caching is permitted.

(a) How much fuel is required if point *B* is 1 unit away?

(b) How much fuel is required if point *B* is 1.005 units away?

(c) How much fuel is required if point *B* is 1.01 units away?

(d) How much fuel is required if point *B* is 1.04 units away?

(e) If you may use only 4.8 tanks of fuel what is the maximum distance of point *B*?

(f) If you may use only 5 tanks of fuel what is the maximum distance of point *B*?

Chapter 10

MathDice Puzzles

In 2004 Sam Ritchie invented the game of MathDice now being sold by ThinkFun, Inc. In MathDice, dice are thrown to determine three scoring numbers (from the set 1 to 6) and a target number to make with a mathematical expression that uses each scoring number once and only once. Each puzzle in this chapter gives three scoring numbers (from the set 0 to 9) and a challenge to make a target number according to certain rules.

Ordered Trigrams

The first sets of problems have rules in which you may use +, −, ×, ÷, exponents, factorials, parentheses and concatenation (that is, combining two digits into another number; for instance, putting a 1 and 2 together to make 21). No roots, decimals or other math functions are permitted. Additionally the digits must appear in ascending order for some puzzles and descending order for others. Also, leading minus signs are not allowed anywhere in the expression.

77
Use 0, 1 and 2 in Ascending Order to Make

(a) 12.
(b) 13.
(c) 24 (2 ways).

78
Use 1, 2 and 3 in Ascending Order to Make

(a) 12 (2 ways).

(b) 13.

(c) 15.

(d) 24 (3 ways).

(e) 27.

(f) 36 (2 ways).

(g) 64 (2 ways).

(h) 72.

79
Use 2, 3 and 4 in Ascending Order to Make

(a) 14 (2 ways).

(b) 16 (3 ways).

(c) 20 (2 ways).

(d) 24 (2 ways).

(e) 30 (3 ways).

(f) 36 (2 ways).

(g) 40.

(h) 47.

(i) 54.

(j) 60 (2 ways).

(k) 68.

(l) 74.

(m) 83.

(n) 88.

(o) 96.

80
Use 3, 4 and 5 in Ascending Order to Make

(a) 12.

(b) 15 (2 ways).

(c) 16.

(d) 17.

(e) 19.

(f) 22.

(g) 24 (3 ways).

(h) 25 (2 ways).

(i) 26.

(j) 27.

(k) 29 (2 ways).

(l) 30.

(m) 32 (2 ways).

(n) 35 (3 ways).

(o) 36.

(p) 39.

(q) 42.

(r) 50.

(s) 51.

(t) 54.

(u) 57.

(v) 60 (2 ways).

(w) 67.

(x) 76.

(y) 77.

(z) 80.

(a1) 86.

(b1) 87.

(c1) 90.

(d1) 99.

81
Use 4, 5 and 6 in Ascending Order to Make

(a) 13.

(b) 20.

(c) 24 (3 ways).

(d) 25.

(e) 26.

(f) 34.

(g) 35.

(h) 39.

(i) 44 (2 ways).

(j) 51.

(k) 54 (2 ways).

(l) 60.

(m) 80 (2 ways).

82
Use 5, 6 and 7 in Ascending Order to Make

(a) 18 (2 ways).

(b) 23.

(c) 24.

(d) 27.

(e) 35.

(f) 37.

(g) 47.

(h) 49.

(i) 53.

(j) 63.

(k) 65.

(l) 72.

(m) 77.

(n) 78.

83
Use 6, 7 and 8 in Ascending Order to Make

(a) 21.

(b) 34.

(c) 48 (2 ways).

(d) 50.

(e) 59.

(f) 62.

(g) 75.

(h) 90 (3 ways).

84
Use 7, 8 and 9 in Ascending Order to Make

(a) 24.

(b) 47.

(c) 63.

(d) 65.

(e) 69.

(f) 70.

(g) 79.

(h) 87.

(i) 96.

85
Use 0, 2 and 4 in Ascending Order to Make

(a) 17.

(b) 23.

(c) 25 (2 ways).

(d) 30 (2 ways).

(e) 49.

(f) 72.

(g) 81.

86
Use 1, 3 and 5 in Ascending Order to Make

(a) 12.

(b) 15.

(c) 16.

(d) 19.

(e) 20 (2 ways).

(f) 29.

(g) 30.

(h) 31.

(i) 40.

(j) 42.

(k) 65.

87
Use 2, 4 and 6 in Ascending Order to Make

(a) 16.

(b) 20 (2 ways).

(c) 22.

(d) 24.

(e) 26 (2 ways).

(f) 30.

(g) 32.

(h) 36 (2 ways).

(i) 42.

(j) 48 (3 ways).

(k) 54.

(l) 56.

(m) 60 (2 ways).

(n) 64.

(o) 96.

88
Use 3, 5 and 7 in Ascending Order to Make

(a) 13.

(b) 15.

(c) 18 (2 ways).

(d) 22.

(e) 23.

(f) 24.

(g) 28.

(h) 36.

(i) 37.

(j) 38.

(k) 41.

(l) 42 (2 ways).

(m) 56.

(n) 60 (2 ways).

(o) 63.

(p) 72.

(q) 77.

89
Use 4, 6 and 8 in Ascending Order to Make

(a) 12.

(b) 16 (2 ways).

(c) 18 (2 ways).

(d) 22.

(e) 26.

(f) 32 (3 ways).

(g) 38 (2 ways).

(h) 52.

(i) 54.

(j) 56.

(k) 72.

(l) 80.

(m) 90.

(n) 92.

(o) 93.

(p) 94.

90
Use 5, 7 and 9 in Ascending Order to Make

(a) 21.

(b) 26.

(c) 41.

(d) 44.

(e) 48.

(f) 57.

(g) 66.

(h) 68.

(i) 80.

(j) 84.

91
Use 0, 3 and 6 in Ascending Order to Make

(a) 13.

(b) 19.

(c) 24 (2 ways).

(d) 30.

(e) 37 (2 ways).

(f) 42.

(g) 64.

92
Use 1, 4 and 7 in Ascending Order to Make

(a) 11 (2 ways).

(b) 12.

(c) 18.

(d) 21.

(e) 24.

(f) 31 (2 ways).

(g) 32.

(h) 35.

(i) 47.

(j) 48.

(k) 98.

93
Use 2, 5 and 8 in Ascending Order to Make

(a) 15.

(b) 17 (2 ways).

(c) 18.

(d) 24.

(e) 26.

(f) 30.

(g) 33.

(h) 40.

(i) 42.

(j) 56.

(k) 60.

(l) 80.

(m) 90.

94
Use 3, 6 and 9 in Ascending Order to Make

(a) 18.

(b) 21.

(c) 24 (2 ways).

(d) 27 (2 ways).

(e) 45 (3 ways).

(f) 48.

(g) 75.

(h) 81 (2 ways).

(i) 83.

(j) 86.

(k) 90.

95
Use 0, 4 and 8 in Ascending Order to Make

(a) 13.

(b) 15.

(c) 17.

(d) 33.

(e) 40.

(f) 48 (2 ways).

(g) 49.

96
Use 1, 5 and 9 in Ascending Order to Make

(a) 15.

(b) 24.

(c) 46.

(d) 54.

(e) 59.

(f) 60.

(g) 80.

97
Use 1, 2 and 3 in Descending Order to Make

(a) 11.

(b) 18.

(c) 23.

(d) 27 (2 ways).

(e) 32 (3 ways).

(f) 35.

(g) 37.

(h) 63.

98
Use 1, 2 and 4 in Descending Order to Make

(a) 11.

(b) 15.

(c) 18.

(d) 21.

(e) 25 (2 ways).

(f) 26 (3 ways).

(g) 30.

(h) 45.

(i) 47.

(j) 64.

99
Use 1, 2 and 5 in Descending Order to Make

(a) 11 (2 ways).

(b) 20.

(c) 24 (2 ways).

(d) 26 (2 ways).

(e) 30.

(f) 40.

(g) 51.

(h) 61.

(i) 99.

100
Use 3, 5 and 6 in Descending Order to Make

(a) 12 (2 ways).

(b) 18.

(c) 21 (2 ways).

(d) 24 (3 ways).

(e) 26.

(f) 33 (2 ways).

(g) 36 (3 ways).

(h) 42.

(i) 46.

(j) 48 (2 ways).

(k) 59 (2 ways).

(l) 66.

(m) 90 (2 ways).

(n) 100.

101
Use 3, 6 and 7 in Descending Order to Make

(a) 13 (2 ways).

(b) 21 (2 ways).

(c) 24 (2 ways).

(d) 39 (2 ways).

(e) 42 (2 ways).

(f) 49.

(g) 70 (2 ways).

(h) 78.

(i) 80.

(j) 82.

(k) 84.

102
Use 3, 6 and 8 in Descending Order to Make

(a) 14.

(b) 24 (2 ways).

(c) 28.

(d) 48.

(e) 50.

(f) 55.

(g) 56.

(h) 57.

(i) 64 (2 ways).

(j) 80.

(k) 96.

103
Use 3, 4 and 7 in Descending Order to Make

(a) 11.

(b) 18 (2 ways).

(c) 25 (2 ways).

(d) 27.

(e) 28 (3 ways).

(f) 31 (3 ways).

(g) 35.
(h) 36.
(i) 37.

(j) 70.
(k) 71 (2 ways).

104
Use 3, 4 and 5 in Descending Order to Make

(a) 11.
(b) 12 (2 ways).
(c) 15 (2 ways).
(d) 16.
(e) 20 (2 ways).
(f) 29 (2 ways).
(g) 30.

(h) 32 (2 ways).
(i) 40 (2 ways).
(j) 51.
(k) 56.
(l) 90 (3 ways).
(m) 96 (2 ways).

105
Use 3, 5 and 8 in Descending Order to Make

(a) 12.
(b) 18 (2 ways).
(c) 24.
(d) 27.
(e) 28.
(f) 32.

(g) 48 (2 ways).
(h) 56.
(i) 64 (2 ways).
(j) 78.
(k) 88 (2 ways).

106
Use 3, 7 and 8 in Descending Order to Make

(a) 14.
(b) 15.
(c) 24 (2 ways).
(d) 29 (2 ways).
(e) 49.

(f) 56.
(g) 63.
(h) 81 (2 ways).
(i) 90 (2 ways).

107
Use 3, 4 and 8 in Descending Order to Make

(a) 12 (3 ways).
(b) 16 (2 ways).
(c) 24 (4 ways).
(d) 28.
(e) 30.
(f) 32 (3 ways).

(g) 35 (2 ways).
(h) 64 (3 ways).
(i) 72 (3 ways).
(j) 80 (2 ways).
(k) 96 (2 ways).

108
Use 3, 4 and 6 in Descending Order to Make

(a) 18.
(b) 27 (3 ways).
(c) 33 (2 ways).
(d) 36 (2 ways).
(e) 40.
(f) 58.

(g) 60 (2 ways).
(h) 64.
(i) 70 (2 ways).
(j) 72 (2 ways).
(k) 90 (2 ways).

109
Use 4, 7 and 9 in Descending Order to Make

(a) 15.
(b) 16.
(c) 18.
(d) 24.
(e) 26.
(f) 39.

(g) 48 (2 ways).
(h) 68.
(i) 73.
(j) 83.
(k) 99.

110
Use 4, 8 and 9 in Descending Order to Make

(a) 13 (2 ways).
(b) 24 (2 ways).
(c) 27.

(d) 33 (2 ways).
(e) 36 (2 ways).
(f) 41 (2 ways).

(g) 68 (2 ways).   (j) 93.
(h) 74.   (k) 96.
(i) 81.

111
Use 3, 4 and 9 in Descending Order to Make

(a) 11 (2 ways).   (h) 40.
(b) 17.   (i) 72 (2 ways).
(c) 20.   (j) 73.
(d) 21 (2 ways).   (k) 88.
(e) 27.   (l) 99.
(f) 33 (3 ways).   (m) 100.
(g) 39 (3 ways).

Trigrams — Intermediate Rules

The next sets of problems have **intermediate rules** in which you may use +, −, ×, ÷, exponents, decimal points, parentheses and concatenation (that is, combining two digits into another number; for instance, putting a 1 and 2 together to make 21). No roots, repeating decimals or other math functions are permitted. Two expressions are considered the same if one can be immediately derived from the other. For example, $1 \div 2^{-3}$ and 1×2^3, $.6 \times 5 - 1$ and $6 \times .5 - 1$ and $42 + 3$ and $43 + 2$ are pairs of equivalent expressions. Bring your calculators and be prepared to consider many possibilities in these exercises.

112
Stimulating Singles

(a) Make **15** using **2, 6** and **8**.   (g) Make **13** using **3, 6** and **7**.
(b) Make **25** using **2, 6** and **9**.   (h) Make **17** using **3, 6** and **9**.
(c) Make **30** using **2, 7** and **9**.   (i) Make **9** using **3, 7** and **8**.
(d) Make **7** using **2, 8** and **8**.   (j) Make **24** using **3, 8** and **8**.
(e) Make **11** using **2, 8** and **9**.   (k) Make **9** using **4, 4** and **4**.
(f) Make **25** using **3, 3** and **4**.   (l) Make **16** using **4, 5** and **5**.

(m) Make **28** using **4, 5** and **5**.
(n) Make **30** using **4, 5** and **8**.
(o) Make **23** using **4, 5** and **9**.
(p) Make **64** using **4, 5** and **9**.
(q) Make **11** using **4, 6** and **8**.
(r) Make **18** using **4, 9** and **9**.

(s) Make **16** using **5, 5** and **5**.
(t) Make **11** using **5, 5** and **6**.
(u) Make **26** using **5, 5** and **6**.
(v) Make **11** using **5, 8** and **9**.
(w) Make **64** using **5, 8** and **9**.

113
Perplexing Pairs

(a) Make **7** using **1, 2** and **2**.
(b) Make **18** using **1, 2** and **2**.
(c) Make **8** using **1, 3** and **3**.
(d) Make **6** using **1, 3** and **4**.
(e) Make **10** using **1, 5** and **8**.
(f) Make **16** using **1, 6** and **9**.
(g) Make **32** using **2, 3** and **9**.
(h) Make **14** using **2, 4** and **5**.
(i) Make **22** using **2, 4** and **9**.
(j) Make **64** using **2, 5** and **5**.
(k) Make **25** using **2, 6** and **8**.
(l) Make **7** using **2, 8** and **9**.
(m) Make **31** using **3, 3** and **4**.

(n) Make **24** using **3, 3** and **5**.
(o) Make **30** using **3, 3** and **9**.
(p) Make **4** using **3, 4** and **5**.
(q) Make **8** using **3, 4** and **9**.
(r) Make **14** using **3, 5** and **8**.
(s) Make **18** using **3, 6** and **6**.
(t) Make **8** using **4, 4** and **5**.
(u) Make **12** using **4, 4** and **5**.
(v) Make **18** using **4, 5** and **5**.
(w) Make **16** using **4, 5** and **6**.
(x) Make **32** using **4, 5** and **8**.
(y) Make **25** using **4, 5** and **9**.

114
Troublesome Trios
(3 Solutions Each)

(a) Make **8** using **1, 1** and **2**.
(b) Make **13** using **1, 2** and **8**.
(c) Make **8** using **1, 3** and **4**.
(d) Make **15** using **2, 2** and **6**.
(e) Make **2** using **2, 3** and **8**.
(f) Make **10** using **2, 3** and **9**.
(g) Make **10** using **2, 5** and **6**.

(h) Make **32** using **2, 5** and **6**.
(i) Make **28** using **2, 5** and **7**.
(j) Make **9** using **3, 3** and **3**.
(k) Make **16** using **3, 4** and **6**.
(l) Make **13** using **3, 5** and **5**.
(m) Make **4** using **3, 5** and **7**.
(n) Make **15** using **3, 5** and **7**.

(o) Make **32** using **3, 5** and **8**.

(p) Make **10** using **4, 5** and **6**.

(q) Make **8** using **4, 5** and **7**.

(r) Make **9** using **4, 5** and **7**.

(s) Make **20** using **4, 5** and **8**.

(t) Make **25** using **5, 5** and **7**.

(u) Make **4** using **5, 6** and **8**.

(v) Make **25** using **5, 7** and **9**.

115
Menacing Multiples
(4–7 Solutions Each)

(a) Make **5** using **2, 6** and **7** (4 ways).

(b) Make **27** using **3, 3** and **6** (4 ways).

(c) Make **32** using **3, 4** and **5** (4 ways).

(d) Make **6** using **3, 5** and **9** (4 ways).

(e) Make **2** using **4, 5** and **8** (4 ways).

(f) Make **8** using **5, 6** and **7** (4 ways).

(g) Make **6** using **5, 7** and **8** (4 ways).

(h) Make **32** using **5, 8** and **8** (4 ways).

(i) Make **64** using **1, 4** and **6** (6 ways).

(j) Make **2** using **2, 4** and **8** (7 ways).

(k) Make **16** using **2, 5** and **5** (5 ways).

(l) Make **32** using **2, 5** and **7** (5 ways).

(m) Make **5** using **2, 5** and **9** (5 ways).

(n) Make **16** using **4, 4** and **8** (5 ways).

116
Largest Possible

(a) Use the digits 1, 2, 3 and 4 once and only once to make a mathematical expression that is as large as possible. You may use +, −, ×, ÷, exponents, decimal points, parentheses and concatenation (that is, combining two digits into another number; for instance, putting a 1 and 2 together to make 21). No roots, factorials, repeating decimals or other math functions are permitted.

(b) Same as above except roots are allowed.

Solutions

Chapter 1: Playful Puzzles

1
Word Mystery

"*What word*" has 8 letters, "*sometimes*" has 9 — "*it always*" contains 8 letters, "*occasionally*" uses 12 though!

Another solution is "*envelope*" in that an envelope can contain any number of written letters inside it and "*envelope*" is an 8 letter word.

2
Salary Secrecy

One solution goes as follows. The first person picks a random number, adds it to his salary, writes the result on a slip of paper and passes that slip to the second person. The second person adds his salary to the number he sees, writes the result on his slip of paper and passes the slip to the third person. The process repeats until the fifth person passes his slip to the first person. The first person subtracts the original random number, divides the result by 5 and announces the result to all as the average salary. To keep salaries confidential the slips of paper are destroyed and all participants must agree not to discuss what numbers they saw with others.

3
Relations Puzzles

(a) The man is pointing to his nephew.

(b) Ray is my great-grandfather.

(c) The speaker and Chris are both female. The speaker is the mother-in-law of Chris.

4
Slider

Move 1 to the lower left corner. Move T to the right. Move 2 to the upper left corner. Move 3 below 2. Move T to the lower right corner.

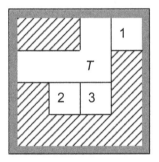

5
Fastest Serve

Using 1 mi. = 1.609344 km, we determine that he served 264.1109127 kph = 164.1109127 mph.

6
The Population Explosion

(a) The surface area of the earth is $A=4\pi R^2$, where $R=6{,}371$ km is the radius of the earth. The volume of a shell of thickness T is $V=AT$. In our case $V_0=0.4599$ km^3 so $T=0.4599/(4\pi R^2)=9.016510^{-4}$ mm.

(b) The time behavior of the volume of human population is $V(t)=V_0\times1.0114^t$, where t is in years. If we equate this to $4\pi R^2T$, where T is 1 meter, we find $V(t)=4\pi R^2/1{,}000$ km$^3=V_0\times1.0114^t$, giving $t=\ln[4\pi R^2/(1000V_0)]/\ln(1.0114)=1{,}227.91$ years. The population at this time is 8.0963 quadrillion.

(c) With $V(t)=V_0\times1.0114^t=(4/3)\times\pi r^3$, we get $r(t)=(0.75\times0.4599/\pi)^{1/3}\times1.0114^{t/3}$. Then $dr(t)/dt=(0.75\times0.4599/\pi)^{1/3}\times\ln(1.0114)\times1.0114^{t/3}/3=9.4605284\times10^{12}$ km per year. From this we determine $t=9578.65$ years. At this time the population is 1.0436×10^{48} people and the radius is 2.5038×10^{15} km $=264.6575$ light years.

7
Catenary

The only way for the conditions to be met is if $d=0$.

Chapter 2: Geometric Puzzles

8
Mining On Rigel IV

(a) Prospector A must have started at any of an infinite set of specific distances each less than 1 mile from the North Pole. This allows him to initially head north and cross over the pole so he may also head north on the third leg of his trek and return to his starting latitude. The smaller circle in Figure (a) is a circle of 1 mile radius centered on the North Pole. Prospector A starts at A and takes the path $ABCDA$.

Let his starting great circle distance from the pole be $AN=s$. Then each of arcs AB, BC, CD and $DAD\cdots k\cdots DA$ must be 1 mile, where arc $DAD\cdots k\cdots DA$ means a path starting at D, going west, circling the North Pole k times and continuing to A. The condition on s for this to happen is

$$2\pi k = \frac{1}{R\sin(\frac{s}{R})} - \frac{1}{R\sin(\frac{1-s}{R})}, \quad \text{where } R=4{,}000 \text{ miles.}$$

Base camp A must be at a distance s_k or $1-s_k$ from the pole for some $k=0, 1, 2, \ldots$. The first few values for s_k and $1-s_k$ are shown in the table.

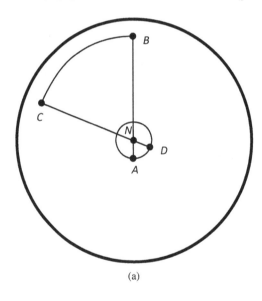

k	s_k	$1-s_k$
0	0.5	0.5
1	0.134435689	0.865564311
2	0.073284499	0.926715501
3	0.050245047	0.949754953
4	0.038208091	0.961791909
5	0.030818801	0.969181199
6	0.025822699	0.974177301
7	0.022219743	0.977780257

(a)

(b) Clearly Prospector B could have started 1/2 mile south of the equator. Base camp B could also be at an infinite number of specific distances, each a little more than 1 mile from the North Pole or each a little less than 1 mile from the South Pole. The larger circle in Figure (b) has a 1 mile radius centered on the South Pole. Prospector B starts at A and takes the path $ABCDA$. Let his starting great circle distance from the pole be r. Then each of arcs AB, BC, CD and $DAD\cdots k\cdots DA$ must be

1 mile, where arc $DAD\cdots k\cdots DA$ means a path starting at D, going west, circling the South Pole k times and continuing to A. The condition on r for this to happen is

$$2\pi k = \frac{1}{R\sin(\frac{r}{R})} - \frac{1}{R\sin(\frac{1-r}{R})}, \quad \text{where } R=4{,}000 \text{ miles.}$$

Base camp B can be at a distance r_k from the South Pole for any $k=1, 2, \ldots$. The first few values for r_k are shown in the table. Base camp B can also be at a distance $1+r_k$ from the North Pole, which can be seen by considering Figure (b) as looking down on the North Pole and taking the path $CDABC$. The first few values for $1+r_k$ in miles are shown in the table.

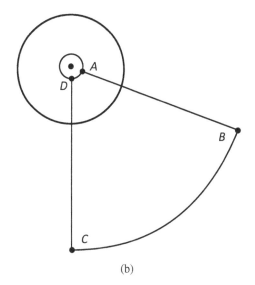

(b)

k	r_k	$1+r_k$
1	0.139652205	1.139652205
2	0.074088383	1.074088383
3	0.050501269	1.050501269
4	0.038320291	1.038320291
5	0.030877565	1.030877565
6	0.025857228	1.025857228
7	0.022241726	1.022241726
8	0.019513588	1.019513588

(c) Figure (c) shows the situation for Prospector C. The North Pole is at the center. He starts at A, $1+r$ from the pole, and goes 1 mile to B. Then he goes east 1 mile, changing his longitude by $\lambda_C=1/[R\cos(\varphi_B)]$ where $\varphi_B=\pi/2-r/R$ is the latitude of B and C. He then goes 1 mile south to D and finally goes 1 mile to E, changing his longitude by

$\Delta\lambda = 1/[R\cos(\varphi_D)]$ where $\varphi_D = \pi/2 - (1+r)/R$ is the latitude of A, D and E. Thus the longitude of E is $\lambda_E = \lambda_C - \Delta\lambda$. The distance from A to E is distance $= 2R \sin^{-1}[\sin(\lambda_E/2)\cos\varphi_B]$. This distance is maximized when base camp C is $1+r = 1.2712313...$ miles from the North Pole. The maximum distance is 2.523974... miles between points A and E.

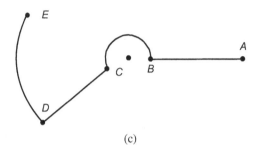

(c)

9
Linking Points

The maximum is 9 links as shown in the figure.

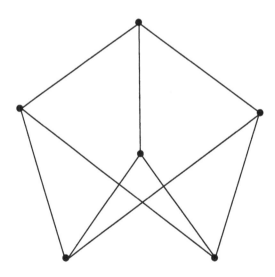

10
Right Triangles

By the Pythagorean Theorem $AB^2 + BC^2 + CD^2 + DE^2 = AC^2 + CD^2 + DE^2 = AD^2 + DE^2 = AE^2 = 111^2 = 12{,}321$.

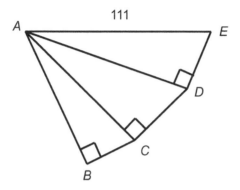

11
The Clipped Polyhedron

When a polyhedron, P_1, with F_1 faces, V_1 vertices and E_1 edges has a small tip sliced off each of its vertices the resulting polyhedron, P_2, will have $F_2 = F_1 + V_1$ faces. Since each vertex of P_2 has three edges meeting there we determine $E_2 = 3V_2/2$. From Euler's relation $E_2 = F_2 + V_2 - 2$. These relations determine $(F_2, V_2, E_2) = (F_1 + V_1, 2E_1, 3E_1)$.

(a) (i) A triangular prism with 3 quadrilateral sides and 2 triangular ends. $(F_1, V_1, E_1) = (5, 6, 9)$. The new polyhedron produced has $(F_2, V_2, E_2) = (11, 18, 27)$.

(ii) A di-pyramid with 6 triangular faces. $(F_1, V_1, E_1) = (6, 5, 9)$. The new polyhedron produced has $(F_2, V_2, E_2) = (11, 18, 27)$. (i) and (ii) are duals of each other.

(b) (i) A quadrilateral with triangular faces at each edge and two more triangular faces to connect the whole. $(F_1, V_1, E_1) = (7, 6, 11)$. The new polyhedron produced has $(F_2, V_2, E_2) = (13, 22, 33)$.

   (ii) Two triangular faces joined at a vertex with 4 quadrilateral faces. $(F_1, V_1, E_1) = (6, 7, 11)$. The new polyhedron produced has $(F_2, V_2, E_2) = (13, 22, 33)$. (i) and (ii) are duals of each other.

  (iii) A pyramid having a quadrilateral base with one of the 4 triangular faces modestly stellated to produce a total of 6 triangular faces. $(F_1, V_1, E_1) = (7, 6, 11)$. The new polyhedron produced has $(F_2, V_2, E_2) = (13, 22, 33)$.

  (iv) A pentagonal base with two quadrilateral faces on sides 1 and 3 and joined with a common edge above the pentagonal base. The remaining 3 faces are triangles. $(F_1, V_1, E_1) = (6, 7, 11)$. The new polyhedron produced has $(F_2, V_2, E_2) = (13, 22, 33)$. (iii) and (iv) are duals of each other.

12
The Papered Boxes

(a) Aksana has a box with dimensions 5, 21 and 210 with a volume of 22,050. The missing face of the box is 5 by 21. Josh has a box with dimensions 12, 12 and 6 with a volume of 864. The missing face is 12 by 12.

(b) Bob has a box with dimensions 3, 6 and 12 with a volume of 216. The missing face is 3 by 12. Kathy has a box with dimensions 3, 6 and 6 with a volume of 108. The missing face is 6 by 6.

(c) Chris has a box with dimensions 5, 36 and 45 with a volume of 8,100. Laurie has a box with dimensions 10, 12 and 15 with a volume of 1,800.

(d) David has a box with dimensions 3, 8 and 24 with a volume of 576. Mary has a box with dimensions 4, 6 and 12 with a volume of 288.

13
The Almost Rectangular Lake

The first figure has regions of the lake identified as $R_1 = EFGM$, $R_2 = BCDM$, $B_1 = AEM$, $B_2 = ADM$ and $B_3 = DEM$. As seen in the second figure $A_1 = DEFG = R_1 + B_3$, $A_2 = BCDE = R_2 + B_3$ and $A_3 = ADE = B_1 + B_2 + B_3$. Note that $B_1 = R_1/2$ and $B_2 = R_2/2$. Thus $A_3 = B_3 + (R_1 + R_2)/2 = (R_1 + B_3)/2 + (R_2 + B_3)/2 = (A_1 + A_2)/2$. Thus $A_3 = (A_1 + A_2)/2$ for an arbitrary shoreline between D and E.

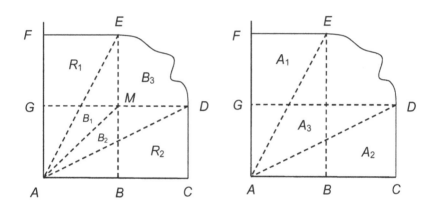

14
The Trifurcated Diamond

The six known solutions are shown below.

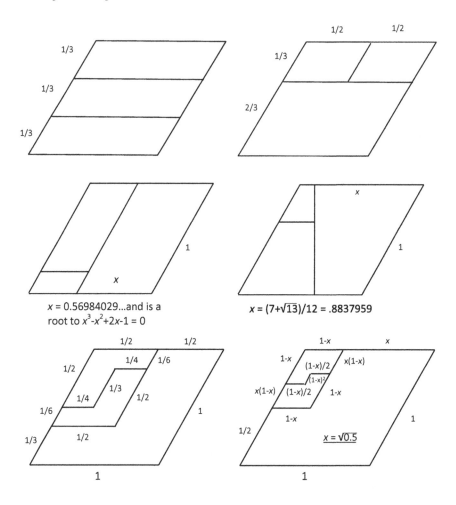

$x = 0.56984029...$and is a
root to $x^3-x^2+2x-1 = 0$

$x = (7+\sqrt{13})/12 = .8837959$

$x = \sqrt{0.5}$

15
Square Dissection

Twelve rectangles in a 96 by 96 square is the smallest number known as is shown in the figure. The smallest rectangle is 1×3.

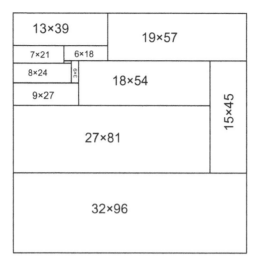

Chapter 3: Digital Puzzles

16
Pandigital Purchases

There were 5 gifts costing as follows: $1+9+25+36+784=855$.

17
Sudoku Variant

18
Pandigital Sums

(a)   (i) $6/7+589/4{,}123=1$.   (ii) $27/39+48/156=1$.   (iii) $3/57+864/912=1$.

(b)   (i) $1/6+7{,}835/9{,}402=1$.

  (ii) $35/70+148/296=38/76+145/290=45/90+138/276=48/96+135/270=1$.

  (iii) $6/534+792/801=4/356+792/801=1$.

(c)   (i) $5/3+916/2{,}748=2$.   (ii) $36/24+79/158=2$.   (iii) $1/37+584/296=2$.

(d)   (i) $4/7+8{,}930/6{,}251=2$.   (ii) $45/39+176/208=2$.   (iii) $1/238+950/476=2$.

19
Even and Odd

(a)   It is not possible. m is odd with digital root 7 while n is even with digital root 2. The digital root of a number is its remainder when divided by 9. This means that if $n=km$ then $k=8$; but the smallest possible m multiplied by 8 has 6 digits.

20
Integer Oddity

I must have the digits 2 and 3. $E_1 = \sqrt[-2]{0.3} = 411.52$, with nearest integer = 412. $E_2 = {}^{-0.3}\!\sqrt[3]{0.2} = 213.75$, with nearest integer = 214.

21
A 10-Digit Number

(a) My number is 9,876,351,240.

(b) My number is 9,123,567,480.

Chapter 4: Logical Puzzles

22
Cake Division

(a) Joe cuts the cake into pieces of size x and $1-x$, where $x \geq 1/2$. If $x > 2/3$ then Bob cuts the larger piece in half and gets $> 1/3$ of the cake. If $x < 2/3$ then Bob divides the smaller piece into 0 and $1-x$ and gets $>1/3$ of the cake. Thus Joe divides the cake into pieces of 2/3 and 1/3, guaranteeing himself 2/3 of the cake.

(b) Joe cuts the cake into pieces of size x and $1-x$, where $x \geq 1/2$. Bob should always cut the smaller piece in half, forcing Joe to take $x + (1-x)/2 = (1+x)/2$ of the cake. Thus Joe should cut the cake into equal halves and Bob cuts one of those in half, forcing Joe to take 3/4 of the cake.

(c) Joe cuts the cake into pieces of size 1/2 and 1/2. Bob will then cut to leave pieces of sizes 1/4, 1/4 and 1/2. Joe then leaves pieces of sizes 1/8, 1/8, 1/4 and 1/2 and thus gets 5/8 of the cake.

23
Prisoner's Escape

The knight overlays the rectangle with 10 3×4 tiles as shown with two tiles overlapping in the center square. He chooses that square to have its number revealed and when it is a 4 he immediately states the total as $2{,}016 = 10 \times 202 - 4$.

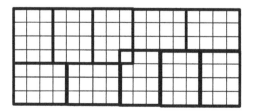

24
Logical Question

The triangle ABC in the figure has a 30° angle at C so the radius of the circle inscribing ABC is $s/(2 \sin 30°)=s$. Thus the answer to the logical question is "No".

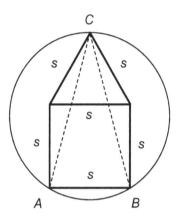

25
Ali Baba and the 10 Thieves

It can be done in 33 crossings. Let the 11 participants be labeled as *AABCDEFGHIJ* in order of descending rank. The table shows the solution. In general if there are n thieves then $4n - 7$ crossings are required.

Move Start	Ranks moved by boat	Ranks on near side All	Ranks on far side None
1	AAB	CDEFGHIJ	AAB
2	AB	ABCDEFGHIJ	A
3	BC	ADEFGHIJ	ABC
4	AB	AABDEFGHIJ	C
5	AAB	DEFGHIJ	AABC
6	BC	BCDEFGHIJ	AA
7	CD	BEFGHIJ	AACD
8	AA	AABEFGHIJ	CD
9	AAB	EFGHIJ	AABCD
10	BC	BCEFGHIJ	AAD
11	EF	BCGHIJ	AADEF
12	AA	AABCGHIJ	DEF
13	AAB	CGHIJ	AABDEF
14	AB	ABCGHIJ	ADEF
15	BC	AGHIJ	ABCDEF
16	AB	AABGHIJ	CDEF
17	AAB	GHIJ	AABCDEF
18	BC	BCGHIJ	AADEF
19	GH	CIJ	AADEFGH
20	AA	AABCIJ	DEFGH
21	AAB	CIJ	AABDEFGH
22	AB	ABCIJ	ADEFGH
23	BC	AIJ	ABCDEFGH
24	AB	AABIJ	CDEFGH
25	AAB	IJ	AABCDEFGH
26	BC	BCIJ	AADEFGH
27	IJ	BC	AADEFGHIJ
28	AA	AABC	DEFGHIJ
29	AAB	C	AABDEFGHIJ
30	AB	ABC	ADEFGHIJ
31	BC	A	ABCDEFGHIJ
32	AB	AAB	CDEFGHIJ
33	AAB	None	All

26
Anniversary Party Puzzles

(a) The children are 4, 4 and 9 so Sum=17 and Product=144. Smith guessed (3, 6, 8) for the ages. One year ago the ages were (3, 3, 8) and Jones guessed (2, 6, 6).

(b) The children are 6, 6 and 28 so Sum=40 and Product=1,008. Smith guessed (3, 16, 21) for the ages. One year ago the ages were (5, 5, 27) and Jones guessed (3, 9, 25). Three years ago the ages were (3, 3, 25) and Brown guessed (1, 15, 15).

(c) The children are 6, 8 and 25 so Sum=39 and Product=1,200. Smith guessed (4, 15, 20) for the ages and Jones guessed (5, 10, 24). Four years ago the ages were (2, 4, 21) and Brown guessed (1, 12, 14).

(d) The children are 9, 10 and 28 so Sum=47 and Product=2,520. Smith guessed (6, 20, 21) for the ages. Four years ago the ages were (5, 6, 24) and Jones guessed (3, 12, 20). Seven years ago the ages were (2, 3, 21) and Brown guessed (1, 7, 18).

(e) The children are 10, 12 and 21 so Sum=43 and Product=2,520. Smith guessed (9, 14, 20) for the ages. Five years ago the ages were (5, 7, 16) and Jones guessed (4, 10, 14). Six years ago the ages were (4, 6, 15) and Brown guessed (3, 10, 12).

(f) The children are 12, 12 and 20 so Sum=44 and Product=2,880. Smith guessed (10, 16, 18) for the ages. Two years ago the ages were (10, 10, 18) and Jones guessed (8, 15, 15). Ten years ago the ages were (2, 2, 10) and Brown guessed (1, 5, 8).

(g) The children are 8, 10 and 27 so Sum=45 and Product=2,160. Smith guessed (6, 15, 24) for the ages. Two years ago the ages were (6, 8, 25) and Jones guessed (4, 15, 20). Six years ago the ages were (2, 4, 21) and Brown guessed (1, 12, 14).

(h) The children are 10, 11 and 24 so Sum=45 and Product=2,640. Smith guessed (8, 15, 22) for the ages. Four years ago the ages were (6, 7, 20) and Jones guessed (4, 14, 15). Eight years ago the ages were (2, 3, 16) and Brown guessed (1, 8, 12).

(i) The children are 6, 8 and 22 so Sum = 36 and Sum of Cubes = 11,376. Smith guessed (1, 15, 20) for the ages. One year ago the ages were (5, 7, 21) and Jones guessed (1, 12, 20).

(j) The children are 9, 9 and 19 so Sum = 37 and Sum of Cubes = 8,317. Smith guessed (5, 16, 16) for the ages. Two years ago the ages were (7, 7, 17) and Jones guessed (3, 13, 15).

(k) The children are 9, 12 and 21 so Sum = 42 and Sum of Cubes = 11,718. Smith guessed (7, 15, 20) for the ages. Three years ago the ages were (6, 9, 18) and Jones guessed (3, 15, 15).

(l) The children are 13, 13 and 23 so Sum = 49 and Sum of Cubes = 16,561. Smith guessed (10, 17, 22) for the ages. Four years ago the ages were (9, 9, 19) and Jones guessed (5, 16, 16).

(m) The children are 11, 19 and 25 so Sum = 55 and Sum of Cubes = 23,815. Smith guessed (10, 22, 23) for the ages. Nine years ago the ages were (2, 10, 16) and Jones guessed (1, 12, 15).

Chapter 5: Probability Puzzles

27
Gambler's Surprise

Each time you win, your stake is multiplied by 1.9; each time you lose, your stake is multiplied by 0.1.

(a) Ending stake is $10,000 \times 1.9^{75} \times 0.1^{25} = \$0.80634$.

(b) Ending stake is $10,000 \times 1.9^{80} \times 0.1^{20} = \$1,996,586.25$.

28
Tenzi

Let $p = 1/6$ and $q = 5/6$. Then if m dice are rolled to achieve a target number the chance of i matches is $P_i = p^i q^{m-i} \times m!/(m-i)!/i!$ Let E_m be the expected number of rolls to achieve 10 matches when m dice don't match the target number. Then $E_m = P_0 E_m + P_1 E_{m-1} + \cdots + P_{m-1} E_1 + 1$. From this we calculate $E_1 = 6$, $E_2 = 96/11$, $E_3 = 10{,}566/1{,}001$, $E_4 = 728{,}256/61{,}061$ and so forth until we get $E_{10} = 98{,}081 \times 336{,}640{,}049 \times 20{,}818{,}956{,}233/(2 \times 3^8 \times 7 \times 11 \times 13 \times 17 \times 29 \times 31 \times 61 \times 113 \times 4{,}651 \times 6{,}959) = 15.3484823859282\ldots$

29
Little Bingo

(a) A winner will always be decided by the 4th drawn number. The first three draws can be one in the N column and two in either the B or G column; any other combination of three draws results in a winning card.

(b) The table below treats the more general case where column B has numbers from 1 to k, column N has numbers from $k+1$ to $2k$ and column G has numbers from $2k+1$ to $3k$. If we add the probabilities in the table then the probability for a row win is $P_{\text{Row}} = 2k(7k-3)/[(9k-3)(3k-2)]$ and the probability of a column win is $P_{\text{Column}} = (13k^2 - 21k + 6)/[(9k-3)(3k-2)]$. For $k = 6$, $P_{\text{Row}} = 117/204 = 0.57353\ldots$ and $P_{\text{Column}} = 87/204 = 0.42647\ldots$

First 2 Turns	Probability	Resolution
BB or *GG*	$2(k-1)/(9k-3)$	Unresolved
BN or *NB*	$2k/(9k-3)$	Unresolved
BG or *GB*	$2k/(9k-3)$	Row Win
NN	$(k-1)/(9k-3)$	Column Win
NG or *GN*	$2k/(9k-3)$	Unresolved
Turn 3	**Probability**	**Resolution**
BBB or *GGG*	$2(k-1)(k-2)/[(9k-3)(3k-2)]$	Column Win
BBN, BNB, …, NGG	$6k(k-1)/[(9k-3)(3k-2)]$	Unresolved
BBG or *GGB*	$2k(k-1)/[(9k-3)(3k-2)]$	Row Win
BNN, NBN, GNN, NGN	$4k(k-1)/[(9k-3)(3k-2)]$	Column Win
BNG, NBG, NGB, GNB	$4k^2/[(9k-3)(3k-2)]$	Row Win
Turn 4	**Probability**	**Resolution**
BBNB, BNBB, …, GGNG	$6k(k-1)(k-2)/[(9k-3)(3k-2)(3k-3)]$	Column Win
BBNN, BNBN, …, GGNN	$6k(k-1)^2/[(9k-3)(3k-2)(3k-3)]$	Column Win
BBNG, BNBG, …, GGNB	$6k^2(k-1)/[(9k-3)(3k-2)(3k-3)]$	Row Win

30
Regular Bingo

(a) A winner will always be decided by the 16th drawn number. The first 15 draws can be *BBBBIIIINNNGGGG* without resolution but the 16th will always produce a winning card.

(b) The analysis for Regular Bingo involves listing every possible distribution of draws and is too lengthy to include here. The result is that $P_{\text{Row}}=0.737342369505\ldots$ and $P_{\text{Column}}=0.252657630494\ldots$, a surprising difference. As with Little Bingo the calculations for P_{Row} and P_{Column} were determined for general k, where $5k$ numbers are drawn from, and the B column has numbers from 1 to k, the I column has numbers from $k+1$ to $2k$, etc. The general expression for P_{Row} is

$P_{\text{Row}}=67.2k^3 P(k)/Q(k)$, where

$P(k)=2{,}427{,}619k^9-40{,}134{,}267k^8+288{,}988{,}538k^7-1{,}186{,}569{,}792k^6$
$+3{,}051{,}795{,}783k^5-5{,}076{,}742{,}911k^4+5{,}428{,}997{,}681k^3-3{,}565{,}395{,}258k^2$
$+1{,}284{,}537{,}834k-187{,}687{,}500.$

$Q(k)=(5k-1)(5k-2)(5k-3)(5k-4)(5k-6)(5k-7)(5k-8)(5k-9)$
$\qquad \times (5k-11)(5k-12)(5k-13)(5k-14).$

For $k=15$ the exact value of P_{Row} is
$924{,}632{,}476{,}308{,}625/1{,}254{,}006{,}977{,}693{,}844.$

31
Fair Duel

Let Brown's probability of hitting Smith be p for each shot and let Smith's probability of winning the duel be S. Then Smith's probability of winning the duel is $S=0.4+0.6(1-p)S$. If we set $S=0.5$ to make the duel fair then $p=2/3$ is Brown's probability of hitting Smith in a single shot.

32
A Golden Set of Tennis

For any single set to be golden one of the two players must win the first 24 points. The probability of this is $p=(1/2)^{23}=1.192092896\ldots\times 10^{-7}$.

(a) Of women's matches played under the conditions of this problem half of them will take 2 sets and half of them will take 3 sets. The probability of no golden set in a 2 set match is $(1-p)^2$ so that the probability of a golden set in a 2 set match is $P_2=2p-p^2$. The probability of no golden set for a 3 set match is $(1-p)^3$ so the probability of a golden set in a 3 set match is $P_3=3p-3p^2+p^3$. The probability of a golden set in a women's match is then $P_w=(P_2+P_3)/2=5p/2-2p^2+p^3/2\approx2.980232\times10^{-7}$.

(b) Of men's matches played under the conditions of this problem 1/4 of them will take 3 sets, 3/8 of them will take 4 sets and 3/8 of them will go 5 sets. The probability of no golden set in a 3 set match is $(1-p)^3$ so that the probability of a golden set in a 3 set match is $P_3=3p-3p^2+p^3$. The probability of no golden set for a 4 set match is $(1-p)^4$ so the probability of a golden set in a 4 set match is $P_4=4p-6p^2+4p^3-p^4$. The probability of no golden set for a 5 set match is $(1-p)^5$ so the probability of a golden set in a 5 set match is $P_5=5p-10p^2+10p^3-5p^4+p^5$. The probability of a golden set in a men's match is then $P_m=(2P_3+3P_4+3P_5)/8=33p/8-27p^2/4+11p^3/2-9p^4/4+3p^5/8\approx4.917383\times10^{-7}$.

33
Bus Ticket Roulette

(a) It is best to bet $2 on an even money bet. Your probability of success is $P=18/38=0.473684\ldots$

(b) It is best to bet any combination of overlapping numbers equivalent to $1 on the top third and simultaneously another $1 on the top half. Immediately after that spin of the wheel you succeed with probability 12/38, you lose with probability 20/38 and your two bets cancel with probability 6/38. Thus $P=12/38+6/38P$, giving $P=3/8=0.375$.

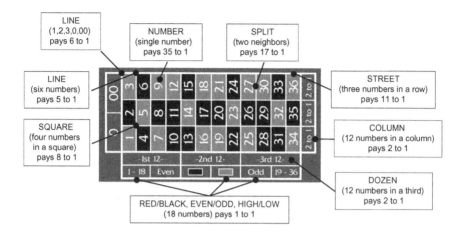

34
Boxes of Colored Balls

(a) We will derive a general expression for e_1, the expected number of balls in the box at the end of the experiment. Let there be n_1 white balls and n_2 black balls initially in the box. Imagine all of them in a line from left to right in order of removal after all of them are removed. For any particular white ball the probability that it will be in the box at the end of the experiment is the probability that all black balls occur in the line to the left of that white ball. This probability is $1/(n_2 + 1)$ and the contribution that this particular white ball makes to e_1 is then $1/(n_2 + 1)$. The same may be said for each of the n_1 white balls so the total contribution to e_1 from white balls is $n_1/(n_2 + 1)$. A similar argument applies to the contributions to e_1 from the black balls giving a net result of $e_1 = n_1/(n_2 + 1) + n_2/(n_1 + 1)$.

(b) We will derive a general expression for e_1, the expected number of balls in the box at the end of the experiment. Let there be n_1 red balls, n_2 white balls and n_3 blue balls initially in the box. Imagine all of them in a line from left to right in order of removal after all of them are removed. For any particular red ball the probability that it will be in the box at the end of the experiment is the probability that all white and blue balls occur in the line to the left of that red ball. This probability is $1/(n_2 + n_3 + 1)$ and the contribution that this particular red ball makes to e_1 is $1/(n_2 + n_3 + 1)$. The same may be said for each of the n_1 red balls so the total contribution to e_1 from red balls is $n_1/(n_2 + n_3 + 1)$. A similar argument applies to the contributions to e_1 from the white and blue balls giving a net result of $e_1 = n_1/(n_2 + n_3 + 1) + n_2/(n_1 + n_3 + 1) + n_3/(n_1 + n_2 + 1)$. A numerical search determined $(n_1, n_2, n_3, e_1) = (172, 597, 17{,}677, 23)$ as the smallest case where e_1 is an integer and n_1, n_2 and n_3 are relatively prime in pairs.

(c) We will derive a general expression for e_2, the expected number of balls in the box at the end of the experiment. Let there be n_1 red balls, n_2 white balls and n_3 blue balls initially in the box. Imagine all of them in a line from left to right in order of removal after all of them are removed. For any particular red ball the probability that it will be in the box at the end of the experiment is the probability that either all white or all blue balls occur in the line to the left of that red ball. This probability is $1/(n_2 + 1) + 1/(n_3 + 1) - 1/(n_2 + n_3 + 1)$ and the contribution that this particular red ball makes to e_2 is $1/(n_2 + 1) + 1/(n_3 + 1) - 1/(n_2 + n_3 + 1)$. The same may be said for each of the n_1 red balls so the total contribution to e_2 from red balls is $n_1/(n_2 + 1) + n_1/(n_3 + 1) - n_1/(n_2 + n_3 + 1)$. A similar argument applies to the contributions to e_2 from the white and blue balls giving a net result of $e_2 = n_1/(n_2 + 1) + n_1/(n_3 + 1) - n_1/(n_2 + n_3 + 1) + n_2/(n_1 + 1) + n_2/(n_3 + 1) - n_2/(n_1 + n_3 + 1) + n_3/(n_1 + 1) + n_3/(n_2 + 1) - n_3/(n_1 + n_2 + 1)$.
A numerical search determined $(n_1, n_2, n_3, e_2) = (4, 9, 25, 8)$ as the only known case where e_2 is an integer.

35
Candyland

The expected number of rolls is $6M + 10$. Consider the more general problem where we use a fair N-sided die and M is much larger than N. For each $k = 0, 1, 2, \ldots, N - 1$ let P_k be the probability that the game ends by landing on square $M + k$. The probability of landing on any specific square $< M$ is very nearly $2/(N + 1)$. The probability of having come from a square prior to M and landing on square $M+k$ is $(N-k)/N$. P_k is the product of these two probabilities: $P_k=2(N-k)/[N(N+1)]$. The sum of kP_k from 1 to N is the expected number of squares of overshoot past M in a game. If S is this sum then $N(N+1)S/2=NS_1-S_2$, where S_1 is the sum of k from 1 to N and S_2 is the sum of k^2 from 1 to N. $NS_1-S_2=N^2(N+1)/2-(2N+1)N(N+1)/6=N(N+1)(N-1)/6$, from which we get $S=(N-1)/3$. The expected number of squares traveled in a game is then $M+S$ and the expected number of rolls per game is $2(M+S)/(N+1)=2M/(N+1)+2(N-1)/[3(N+1)]$. For our particular problem $N=6$ and the average number of rolls per game is $2M/7+10/21$. In 21 games the expected number of rolls is $6M+10$.

Chapter 6: Analytical Puzzles

36
Airport Rush

You should tie your shoe on the walkway. Off the walkway you spend the minute it takes you to tie your shoe making no progress. On the walkway you spend the minute it takes you to tie your shoe making the progress of the walkway. This can be seen more clearly if we imagine one case where you tie your shoe an instant before you would get on the walkway and compare that to the case where you tie your shoe an instant after you get on the walkway.

37
Which Meal?

He is enjoying lunch. The conditions are met only when the hour hand is on the 11th second mark and the minute hand is on the 12th second mark. Thus it is either 2:12 am or 2:12 pm. Since one doesn't normally eat a meal at 2:12 am Kevin must be eating lunch.

38
Rat Race

The tetrahedron can be mapped as shown in the figure. The fast cat can travel around the "square" at a pace slightly exceeding the rat's top speed. The other two cats oscillate along the cross pieces at half the fast cat's speed in such a way that they reach the vertices simultaneously with the fast cat. The rat will have no escape from this strategy.

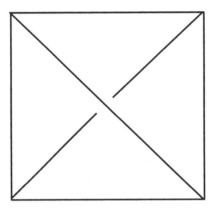

39
Family Visits

(a) The figure shows how to generally solve this problem. Charles starts on foot at velocity v_1. Adam carries Bill to point B at velocity v_3, drops him off, and Bill continues on foot to Grandpa's at velocity v_2. Adam then pedals back to point C at velocity, v_4, to pick up Charles and both ride to Grandpa's, arriving at the same time as Bill. From the figure we note

$$v_1 t_2 = y_1$$

$$v_3 t_1 = y_2$$

$$v_2(t_3 - t_1) = y_3 - y_2$$

$$v_3(t_3 - t_2) = y_3 - y_1$$

$$v_4(t_2 - t_1) = y_2 - y_1.$$

From these equations we eliminate the unknowns, y_1, y_2, t_1, and t_2 to get the distance to Grandpa's as $y_3 = t_3[v_3(v_1 + v_2) + v_4(v_2 + v_3)]/(v_1 + v_3 + 2v_4)$. They get farther if the slower brother goes on foot first. In our case with $(v_1, v_2, v_3, v_4, t_3) = (9, 13, 30, 40, 3)$ we get 60 miles for the distance to Grandpa's.

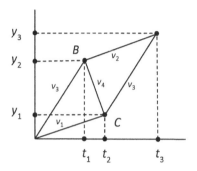

(b) Using the relation $y_3 = t_3[v_3(v_1 + v_2) + v_4(v_2 + v_3)]/(v_1 + v_3 + 2v_4)$ we find Team A with $v_1 = 5$ and $v_2 = 10$ gets to $y_3 = 42$ miles in 3 hours. Team B with $v_1 = 10$ and $v_2 = 5$ gets to $y_3 = 42$ miles in $3\frac{117}{185}$ hours and thus will be $37\frac{35}{37}$ minutes late to Aunt Jenny's.

40
Logging Problem

Define x as $x = \log_{16} 7 \times \log_{49} 625$. Let $p = \log_{16} 7$ and $q = \log_{49} 625$. We know $16^p = 7$ so that $\log_{10} 7 = p \log_{10} 16 = 4p \log_{10} 2$. We also know $49^q = 625$ so that $q \log_{10} 49 = \log_{10} 625$, which simplifies to $2q \log_{10} 7 = 4 \log_{10} 5$. By using $\log_{10} 5 = 1 - \log_{10} 2$ we get $\log_{10} 2 = 1/(1 + 2pq) = 1/(1 + 2x)$.

41
Polynomial Problem 1

$P(x, y) = a_0 + a_1 x + a_2 y + a_3 x^2 + a_4 xy + a_5 y^2 + a_6 x^3 + a_7 x^2 y + a_8 xy^2 + a_9 y^3$.
From $P(0, 0)$ we conclude $a_0 = 0$.
From $P(1, 0) = P(-1, 0) = 0$ we conclude $a_3 = 0$.

From $P(0, 1) = P(0, -1) = 0$ we conclude $a_5 = 0$.

At this stage $P(x, y) = a_1x + a_2y + a_4xy + a_6x^3 + a_7x^2y + a_8xy^2 + a_9y^3$.

From $P(1, 1) + P(1, -1) = 0$ we conclude $a_1 + a_6 + a_8 = 0$.

From $P(1, 0) = 0$ we conclude $a_1 + a_6 = 0$, so it follows $a_8 = 0$ and $a_6 = -a_1$.

From $P(0, 1) = 0$ we conclude $a_2 + a_9 = 0$, so $a_9 = -a_2$.

From $P(1, -1) = 0$ we conclude $a_2 + a_4 + a_7 + a_9 = 0$ and thus $a_7 = -a_4$.

At this stage $P(x, y) = a_1(x - x^3) + a_2(y - y^3) + a_4(xy - x^2y)$.

Suppose $P(k, k) = 0$ (In our problem $k = 12$.). Then $P(x, y) = a_1(k - k^3) + a_2(k - k^3) + a_4(k^2 - k^3) = 0$.

From $P(k, k) = 0$, $P(x, y) = a_1(k - k^3) + a_2(k - k^3) + a_4(k^2 - k^3) = 0$ and $a_4 = -(k + 1/k)(a_1 + a_2) = r(a_1 + a_2)$.

If $a_2 = 0$ we conclude $y = (1 + x)/r$. If $a_1 = 0$ we conclude $1 - y^2 = rx(1 - x)$. These two equations lead to $(r + 1/r^2)x^2 + (2/r^2 - r)x + 1/r^2 - 1 = 0$, with solutions $x = (r^2 + 1)/(r^2 - r + 1)$; $y = (1 - 2r)/(r^2 - r + 1)$.

In terms of k $x = (2k + 1)/(3k^2 + 3k + 1)$ and $y = (3k^2 + 2k)/(3k^2 + 3k + 1)$.

For $k = 12$ we get $x = 25/469$ and $y = 456/469$.

42
Polynomial Problem 2

(a) Write the equation generally as $xy + ax + by + c = dx^3$, where a, b, c and d are integer constants. Solve for y giving $y = dx^2 - bdx + b^2d - a + (ab - b^3d - c)/(x + b)$. In our problem the last term is $-1,008/(x + 5)$. For this to be an integer $x + 5$ must be $\pm$ a divisor of 1,008. There are 30 positive divisors of 1,008 so there are 60 integer solutions.

(b) There are 7 solutions where both x and y are primes. $(x, y) = (2, 7)$, $(7, 227)$, $(31, 6,619)$, $(-3, -113)$, $(-19, 3,919)$, $(-23, 5,407)$, $(-89, 67,134)$, and $(-173, 246,557)$.

(c) There are 4 solutions where both x and y are negative. $(x, y) = (-1, -5)$, $(-2, -25)$, $(-3, -113)$, and $(-4, -521)$.

43
Cascaded Prime Triangles

For x_0, y_0 and x_1 to be the sides of a Pythagorean triangle the most general solution is $x_0 = K(m^2 - n^2)$, $y_0 = 2Kmn$ and $x_1 = K(m^2 + n^2)$, where m and n are of opposite parity and have no common factor. Since we require that x_0 and x_1 are primes it follows that $K=1$ and $n=m-1$. From this we get $x_0 = 2m-1$; $x_1 = (x_0^2 + 1)/2$; $x_2 = (x_1^2 + 1)/2$; $x_3 = (x_2^2 + 1)/2$; $x_4 = (x_3^2 + 1)/2$ and so on.

(a) The first several examples with three cascaded Pythagorean triangles have $(m, x_0) = (136, 271)$, $(175, 349)$, $(1,501, 3,001)$, $(5,050, 10,099)$, $(5,860, 11,719)$,...

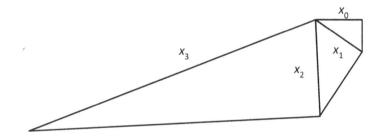

(b) The first several examples with four cascaded Pythagorean triangles have $(m, x_0) = (84,610, 169,219)$, $(685,135, 1,370,269)$, $(2,982,850, 5,965,699)$, $(7,613,940, 15,227,879)$, $(8,875,491, 17,750,981)$, $(9,671,280, 19,342,559)$,...

44
Egg Farming

(a) Eight of Abel's hens will lay $7 \times (7/6)$ eggs in $6/5$ days for a rate of $(7/8) \times (7/6) \div (6/5) = 245/288$ eggs per hen per day. Six of Barry's hens will lay $5 \times (7/6)$ eggs in $8/7$ days for a rate of $(5/6) \times (7/6) \div (8/7) = 245/288$ eggs per hen per day. Thus their hens are equally productive.

(b) With 48 hens Abel gets $245/6$ eggs daily so must wait 6 days to get 245 eggs.

(c) With 300 hens Barry gets 6,125/24 eggs daily so must wait 24 days to get 6,125 eggs.

45
The Knight's Dilemma

The knight must walk a regular polygon of n sides. Its area is $A=nsh/2$, where s is the side length and h is the distance from the center of the polygon to the center of a side $[2h/s=\cot(180°/n)]$. The time in minutes to traverse the polygon is ns/v for the n sides and $kn/60$ for pounding in the n stakes where each stake takes k seconds. These two times add to 24 hr $=1,440$ min so that $ns/v=1,440-kn/60$. To maximize the area pick k and then find that n which maximizes
$A=v^2\cot(180°/n)(1,440-nk/60)^2/(4n)$. For example, when k is 10, $n=31$ and when k is 60, $n=17$. By checking other k it so happens that when k is 23 n is 23, and the area is largest, for that case with $A=161,982.4505v^2$ $=1.61982\times10^9$ feet$^2=58.103209$ square miles.

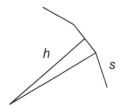

46
The Least Equilateral Triangle

Scale the figure so that $AB=1$.

From the sine law in triangle ABC, $BC=\sin A/\sin C$ and $AC=\sin B/\sin C$.

From the sine law in triangle ABC', $AC'=AB\ \sin(\beta)/\sin(\alpha+\beta)=\sin(\beta)/\sin(\alpha+\beta)$.

From the sine law in triangle $A'BC$, $A'B=BC\ \sin(\gamma)/\sin(\beta+\gamma)=\sin A\ \sin(\gamma)/\sin C/\sin(\beta+\gamma)$.

From the sine law in triangle $AB'C$, $AB'=AC\ \sin(\gamma)/\sin(\alpha+\gamma)=\sin B\ \sin(\gamma)/\sin C/\sin(\alpha+\gamma)$.

From these expressions the coordinates of A', B' and C' are

$C'_x = AC' \cos(\alpha) = \sin(\beta) \cos(\alpha)/\sin(\alpha+\beta);$

$C'_y = AC' \sin(\alpha) = \sin(\beta) \sin(\alpha)/\sin(\alpha+\beta).$

$A'_x = 1 - A'B \cos(B-\beta) = 1 - \sin A \sin(\gamma) \cos(B-\beta)/\sin C/\sin(\beta+\gamma);$

$A'_y = A'B \sin(B-\beta) = \sin A \sin(\gamma) \sin(B-\beta)/\sin C/\sin(\beta+\gamma).$

$B'_x = AB' \cos(A-\alpha) = \sin B \sin(\gamma) \cos(A-\alpha)/\sin C/\sin(\alpha+\gamma);$

$B'_y = AB' \sin(A-\alpha) = \sin B \sin(\gamma) \sin(A-\alpha)/\sin C/\sin(\alpha+\gamma).$

A numerical review of the problem establishes that we can get arbitrarily close to the upper bound for f if we let $A = 180° - 2\varepsilon$, $B = \varepsilon$ and $C = \varepsilon$, where ε is arbitrarily small and we let n become arbitrarily large. With these choices $B'C' = a' = 2\varepsilon/\pi$, $A'C' = b' = A'B' = c' = \varepsilon\sqrt{(1+1/\pi^2)}$. This leads to the angles of the limiting triangle $A'B'C'$ as shown in the figure, where $\varphi = 2a \tan(1/\pi) = 35.3135743°$, $\theta = 72.34321285°$ and the upper bound to f is $74.05927709°$.

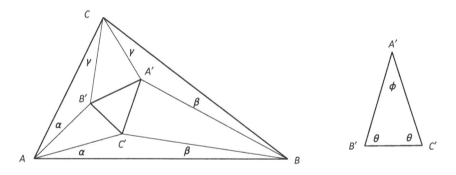

47

You're the Doctor

Notation: (A, B, fA, fB) indicates A ml and fA of a tablet in the larger vial and B ml and fB of a tablet in the smaller vial. The first step always begins by filling one or other vial and proceeds in sequence after each step. Putting in the tablet counts as a step.

(a) **Capacities of 5 ml and 4 ml.** There are solutions for $n=1$–10, 12–21, 24–28, 30, 32–36, 38–40, 45, 48–52, 56, 57, 60, 61, 64–68, 70, 74–76, 80 and 85.

$f=$ **20% or 80%** (3 steps): (5, 0, 0, 0), (5, 0, 1, 0), (1, 4, 0.2, 0.8)

$f=$ **25% or 75%** (5 steps): (0, 4, 0, 0), (4, 0, 0, 0), (4, 4, 0, 0), (4, 4, 0, 1), (5, 3, 0.25, 0.75)

$f=$ **4% or 16%** (6 steps): (1, 4, 0.2, 0.8), (1, 0, 0.2, 0), (5, 0, 0.2, 0), (1, 4, 0.04, 0.16)

$f=$ **64%** (7 steps): (1, 4, 0.2, 0.8), (0, 4, 0, 0.8), (4, 0, 0.8, 0), (5, 0, 0.8, 0), (1, 4, 0.16, 0.64)

$f=$ **15% or 85%** (7 steps): (4, 0, 0, 0), (4, 0, 1, 0), (4, 4, 1, 0), (5, 3, 1, 0), (4, 4, 0.8, 0.2), (5, 3, 0.85, 0.15)

$f=$ **5%** (7 steps): (5, 3, 0.25, 0.75), (5, 0, 0.25, 0), (1, 4, 0.05, 0.2)

$f=$ **40% or 60%** (7 steps): (5, 0, 0, 0), (1, 4, 0, 0), (1, 0, 0, 0), (0, 1, 0, 0), (5, 1, 0, 0), (5, 1, 1, 0), (2, 4, 0.4, 0.6)

$f=$ **32% or 68%** (8 steps): (5, 3, 0.85, 0.15), (4, 4, 0.68, 0.32)

$f=$ **12%** (9 steps): (4, 0, 0.8, 0), (4, 4, 0.8, 0), (5, 3, 0.8, 0), (4, 4, 0.64, 0.16), (5, 3, 0.68, 0.12)

$f=$ **6% or 14%** (9 steps): (1, 0, 0.2, 0), (0, 1, 0, 0.2), (5, 1, 0, 0.2), (2, 4, 0, 0.2), (5, 1, 0.15, 0.05), (2, 4, 0.06, 0.14)

$f=$ **17%** (9 steps): (5, 3, 0.85, 0.15), (5, 0, 0.85, 0), (1, 4, 0.17, 0.68)

$f=$ **24% or 76%** (9 steps): (4, 4, 0.68, 0.32), (5, 3, 0.76, 0.24)

$f=$ **30% or 70%** (9 steps): (0, 1, 0, 0), (0, 1, 0, 1), (5, 1, 0, 1), (2, 4, 0, 1), (5, 1, 0.75, 0.25), (2, 4, 0.3, 0.7)

$f=$ **34% or 66%** (9 steps): (2, 4, 0.4, 0.6), (5, 1, 0.85, 0.15), (2, 4, 0.34, 0.66)

$f=$ **8%** (9 steps): (5, 3, 0.25, 0.75), (4, 4, 0.2, 0.8), (5, 3, 0.4, 0.6), (5, 0, 0.4, 0), (1, 4, 0.08, 0.32)

$f=$ **49% or 51%** (9 steps): (5, 3, 0.4, 0.6), (4, 4, 0.32, 0.68), (5, 3, 0.49, 0.51)

$f=$ **50%** (9 steps): (4, 4, 0, 0), (5, 3, 0, 0), (0, 3, 0, 0), (3, 0, 0, 0), (3, 4, 0, 0), (3, 4, 0, 1), (5, 2, 0.5, 0.5)

$f=$ **3%** (10 steps): (5, 1, 0.15, 0.05), (5, 0, 0.15, 0), (1, 4, 0.03, 0.12)

$f=$ **45%** (10 steps): (5, 3, 0.25, 0.75), (0, 3, 0, 0.75), (3, 0, 0.75, 0), (3, 4, 0.75, 0), (5, 2, 0.75, 0), (3, 4, 0.45, 0.3)

$f=$ **1%** (10 steps): (1, 4, 0.05, 0.2), (1, 0, 0.05, 0), (5, 0, 0.05, 0), (1, 4, 0.01, 0.04)

f=**33% or 67%** (11 steps): (2, 4, 0.3, 0.7), (5, 1, 0.825, 0.175),
(2, 4, 0.33, 0.67)

f=**48%** (11 steps): (2, 4, 0.4, 0.6), (0, 4, 0, 0.6), (4, 0, 0.6, 0), (5, 0, 0.6, 0),
(1, 4, 0.12, 0.48)

f=**10%** (11 steps): (5, 2, 0.5, 0.5), (5, 0, 0.5, 0), (1, 4, 0.1, 0.4)

f=**65% or 35%** (11 steps): (5, 2, 0.5, 0.5), (3, 4, 0.3, 0.7), (5, 2, 0.65, 0.35)

f=**9%** (12 steps): (5, 3, 0.85, 0.15), (0, 3, 0, 0.15), (3, 0, 0.15, 0),
(3, 4, 0.15, 0), (5, 2, 0.15, 0), (3, 4, 0.09, 0.06)

f=**36% or 39%** (12 steps): (3, 4, 0.45, 0.3), (5, 2, 0.6, 0.15),
(3, 4, 0.36, 0.39)

f=**52%** (12 steps): (3, 4, 0, 0), (5, 2, 0, 0), (5, 2, 1, 0), (3, 4, 0.6, 0.4),
(5, 2, 0.8, 0.2), (3, 4, 0.48, 0.52)

f=**61%** (12 steps): (5, 2, 0.65, 0.35), (3, 4, 0.39, 0.61)

f=**18%** (13 steps): (5, 3, 0.76, 0.24), (0, 3, 0, 0.24), (5, 3, 0, 0.24),
(4, 4, 0, 0.24), (5, 3, 0.06, 0.18)

f=**56%** (13 steps): (2, 4, 0.3, 0.7), (0, 4, 0, 0.7), (4, 0, 0.7, 0), (5, 0, 0.7, 0),
(1, 4, 0.14, 0.56)

f=**28%** (13 steps): (2, 4, 0.4, 0.6), (2, 0, 0.4, 0), (0, 2, 0, 0.4), (5, 2, 0, 0.4),
(3, 4, 0, 0.4), (5, 2, 0.2, 0.2), (3, 4, 0.12, 0.28)

f=**26% or 74%** (13 steps): (3, 4, 0.48, 0.52), (5, 2, 0.74, 0.26)

f=**13%** (13 steps): (5, 2, 0.65, 0.35), (5, 0, 0.65, 0), (1, 4, 0.13, 0.52)

f=**57%** (14 steps): (0, 3, 0, 0.75), (0, 4, 0, 0.75), (4, 0, 0.75, 0),
(4, 4, 0.75, 0), (5, 3, 0.75, 0), (4, 4, 0.6, 0.15),
(5, 3, 51/80, 9/80), (4, 4, 0.51, 0.24), (5, 3, 0.57, 0.18)

f=**27%** (14 steps): (3, 4, 0.45, 0.3), (3, 0, 0.45, 0), (3, 4, 0.45, 0),
(5, 2, 0.45, 0), (3, 4, 0.27, 0.18)

f=**2%** (14 steps): (1, 4, 0.08, 0.32), (1, 0, 0.08, 0), (0, 1, 0, 0.08),
(5, 1, 0, 0.08), (2, 4, 0, 0.08), (5, 1, 0.06, 0.02)

f=**21%** (15 steps): (2, 4, 0.3, 0.7), (2, 0, 0.3, 0), (0, 2, 0, 0.3), (5, 2, 0, 0.3),
(3, 4, 0, 0.3), (5, 2, 0.15, 0.15), (3, 4, 0.09, 0.21)

f=**7%** (15 steps): (5, 2, 0.5, 0.5), (3, 4, 0.3, 0.7), (5, 2, 0.65, 0.35),
(0, 2, 0, 0.35), (2, 0, 0.35, 0), (5, 0, 0.35, 0), (1, 4, 0.07, 0.28)

f=**19%** (17 steps): (5, 1, 0.75, 0.25), (0, 1, 0, 0.25), (0, 4, 0, 0.25),
(4, 0, 0.25, 0), (4, 4, 0.25, 0), (5, 3, 0.25, 0),
(4, 4, 0.2, 0.05), (5, 3, 17/80, 3/80), (4, 4, 0.17, 0.08),
(5, 3, 0.19, 0.06)

$f=38\%$ (18 steps): (5, 2, 0.5, 0.5), (0, 2, 0, 0.5), (0, 4, 0, 0.5),
(4, 0, 0.5, 0), (4, 4, 0.5, 0), (5, 3, 0.5, 0),
(4, 4, 0.4, 0.1), (5, 3, 0.425, 0.075),
(4, 4, 0.34, 0.16), (5, 3, 0.38, 0.12)

(b) **Capacities of 5 ml and 1 ml.** There are solutions for $n=1$–6, 8–10, 12, 15, 16, 18, 20, 24, 25, 27, 30, 32, 36, 40, 45, 48, 50, 60, 64, 75 and 80.

$f=20\%$ **or** 80% (3 steps): (5, 0, 0, 0), (5, 0, 1, 0), (4, 1, 0.8, 0.2)
$f=60\%$ (5 steps): (4, 1, 0.8, 0.2), (3, 2, 0.6, 0.4)
$f=25\%$ **or** 75% (5 steps): (5, 0, 0, 0), (4, 1, 0, 0), (4, 0, 0, 0), (4, 0, 1, 0),
(3, 1, 0.75, 0.25)
$f=16\%$ **or** 64% (6 steps): (4, 1, 0.8, 0.2), (4, 0, 0.8, 0), (5, 0, 0.8, 0),
(4, 1, 0.64, 0.16)
$f=50\%$ (6 steps): (0, 1, 0, 0), (1, 0, 0, 0), (1, 1, 0, 0), (2, 0, 0, 0),
(2, 0, 1, 0), (1, 1, 0.5, 0.5)
$f=4\%$ (7 steps): (4, 1, 0.8, 0.2), (0, 1, 0, 0.2), (1, 0, 0.2, 0), (5, 0, 0.2, 0),
(4, 1, 0.16, 0.04)
$f=12\%$ **or** 48% (8 steps): (3, 1, 0.6, 0.2), (3, 0, 0.6, 0), (5, 0, 0.6, 0),
(4, 1, 0.48, 0.12)
$f=10\%$ (8 steps): (0, 1, 0, 0.2), (1, 0, 0.2, 0), (1, 1, 0.2, 0), (2, 0, 0.2, 0),
(1, 1, 0.1, 0.1)
$f=15\%$ (8 steps): (3, 1, 0.75, 0.25), (3, 0, 0.75, 0), (5, 0, 0.75, 0),
(4, 1, 0.6, 0.15)
$f=45\%$ (9 steps): (3, 1, 0.6, 0.2), (3, 0, 0.6, 0), (3, 1, 0.6, 0), (4, 0, 0.6, 0),
(3, 1, 0.45, 0.15)
$f=5\%$ (9 steps): (3, 1, 0.75, 0.25), (0, 1, 0, 0.25), (1, 0, 0.25, 0), (5, 0, 0.25, 0),
(4, 1, 0.2, 0.05)
$f=8\%$ **or** 32% (10 steps): (3, 0, 0.6, 0), (2, 1, 0.4, 0.2), (2, 0, 0.4, 0),
(5, 0, 0.4, 0), (4, 1, 0.32, 0.08)
$f=36\%$ (10 steps): (4, 1, 0.48, 0.12), (4, 0, 0.48, 0), (3, 1, 0.36, 0.12)
$f=2\%$ (11 steps): (1, 1, 0.1, 0.1), (1, 0, 0.1, 0), (5, 0, 0.1, 0),
(4, 1, 0.08, 0.02)
$f=30\%$ (11 steps): (3, 1, 0.45, 0.15), (3, 0, 0.45, 0), (2, 1, 0.3, 0.15)
$f=24\%$ (12 steps): (3, 1, 0.36, 0.12), (3, 0, 0.36, 0), (2, 1, 0.24, 0.12)
$f=9\%$ (12 steps): (3, 0, 0.45, 0), (5, 0, 0.45, 0), (4, 1, 0.36, 0.09)
$f=3\%$ (12 steps): (4, 1, 0.6, 0.15), (0, 1, 0, 0.15), (1, 0, 0.15, 0),
(5, 0, 0.15, 0), (4, 1, 0.12, 0.03)

f=**6%** (13 steps): (4, 1, 0.08, 0.02), (4, 0, 0.08, 0), (3, 1, 0.06, 0.02)

f=**1%** (13 steps): (4, 1, 0.2, 0.05), (0, 1, 0, 0.05), (1, 0, 0.05, 0),
(5, 0, 0.05, 0), (4, 1, 0.04, 0.01)

f=**27%** (14 steps): (3, 0, 0.36, 0), (3, 1, 0.36, 0), (4, 0, 0.36, 0),
(3, 1, 0.27, 0.09)

f=**18%** (16 steps): (3, 1, 0.27, 0.09), (3, 0, 0.27, 0), (2, 1, 0.18, 0.09)

(c) **Capacities of 9 ml, and 1 ml.** There are solutions for n=1–10, 12, 14–16, 18, 20, 21, 24, 25, 27, 28, 30, 32, 35, 36, 40, 42, 45, 48–50, 56, 60, 64, 70, 75 and 80.

f=**50%** (6 steps): (0, 1, 0, 0), (1, 0, 0, 0), (1, 1, 0, 0), (2, 0, 0, 0), (2, 0, 1, 0),
(1, 1, 0.5, 0.5)

f=**75%** (7 steps): (9, 0, 0, 0), (8, 1, 0, 0), (8, 0, 0, 0), (8, 0, 1, 0),
(7, 1, 0.875, 0.125), (7, 0, 0.875, 0), (6, 1, 0.75, 0.125)

f=**25%** (10 steps): (2, 0, 0, 0), (2, 1, 0, 0), (3, 0, 0, 0), (3, 1, 0, 0),
(4, 0, 0, 0), (4, 0, 1, 0), (3, 1, 0.75, 0.25)

f=**20% or 80%** (11 steps): (8, 0, 0, 0), (7, 1, 0, 0), (7, 0, 0, 0), (6, 1, 0, 0),
(6, 0, 0, 0), (5, 1, 0, 0), (5, 0, 0, 0), (5, 0, 1, 0),
(4, 1, 0.8, 0.2)

f=**60%** (13 steps): (4, 1, 0.8, 0.2), (4, 0, 0.8, 0), (3, 1, 0.6, 0.2)

f=**10% or 40%** (15 steps): (6, 1, 0.75, 0.125), (6, 0, 0.75, 0),
(5, 1, 0.625, 0.125), (5, 0, 0.625, 0),
(4, 1, 0.5, 0.125), (4, 0, 0.5, 0), (4, 1, 0.5, 0),
(5, 0, 0.5, 0), (4, 1, 0.4, 0.1)

f=**16% or 64%** (15 steps): (4, 0, 0.8, 0), (4, 1, 0.8, 0), (5, 0, 0.8, 0),
(4, 1, 0.64, 0.16)

f=**15%** (16 steps): (3, 1, 0.75, 0.25), (3, 0, 0.75, 0), (3, 1, 0.75, 0),
(4, 0, 0.75, 0), (4, 1, 0.75, 0), (5, 0, 0.75, 0),
(4, 1, 0.6, 0.15)

f=**30%** (17 steps): (4, 1, 0.4, 0.1), (4, 0, 0.4, 0), (3, 1, 0.3, 0.1)

f=**48%** (17 steps): (4, 1, 0.64, 0.16), (4, 0, 0.64, 0), (3, 1, 0.48, 0.16)

f=**45%** (17 steps): (3, 1, 0.6, 0.2), (3, 0, 0.6, 0), (3, 1, 0.6, 0),
(4, 0, 0.6, 0), (3, 1, 0.45, 0.15)

f=**35%** (19 steps): (7, 0, 0.875, 0), (7, 1, 0.875, 0), (8, 0, 0.875, 0),
(7, 1, 49/64, 7/64), (7, 0, 49/64, 0),
(6, 1, 21/32, 7/64), (6, 0, 21/32, 0),
(5, 1, 35/64, 7/64), (5, 0, 35/64, 0),

(4, 1, 7/16, 7/64), (4, 0, 7/16, 0), (4, 1, 7/16, 0),
(5, 0, 7/16, 0), (4, 1, 0.35, 0.0875)

f=**8% or 32%** (19 steps): (4, 0, 0.4, 0), (4, 1, 0.4, 0), (5, 0, 0.4, 0),
(4, 1, 0.32, 0.08)

f=**12%** (19 steps): (4, 0, 0.6, 0), (4, 1, 0.6, 0), (5, 0, 0.6, 0), (4, 1, 0.48, 0.12)

f=**5%** (20 steps): (9, 0, 0, 0), (9, 0, 1, 0), (8, 1, 8/9, 1/9), (0, 1, 0, 1/9),
(1, 0, 1/9, 0), (1, 1, 1/9, 0), (2, 0, 1/9, 0), (2, 1, 1/9, 0),
(3, 0, 1/9, 0), (3, 1, 1/9, 0), (4, 0, 1/9, 0), (4, 1, 1/9, 0),
(5, 0, 1/9, 0), (4, 1, 4/45, 1/45), (4, 0, 4/45, 0),
(3, 1, 1/15, 1/45), (3, 0, 1/15, 0), (3, 1, 1/15, 0),
(4, 0, 1/15, 0), (3, 1, 0.05, 1/60)

f=**2%** (20 steps): (7, 1, 0.875, 0.125), (0, 1, 0, 0.125), (1, 0, 0.125, 0),
(1, 1, 0.125, 0), (2, 0, 0.125, 0), (2, 1, 0.125, 0),
(3, 0, 0.125, 0), (3, 1, 0.125, 0), (4, 0, 0.125, 0),
(4, 1, 0.125, 0), (5, 0, 0.125, 0), (4, 1, 0.1, 0.025),
(4, 0, 0.1, 0), (4, 1, 0.1, 0), (5, 0, 0.1, 0), (4, 1, 0.08, 0.02)

f=**24%** 21 steps): (4, 1, 0.32, 0.08), (4, 0, 0.32, 0), (3, 1, 0.24, 0.08)

f=**70%** (21 steps): (5, 0, 0.8, 0), (5, 1, 0.8, 0), (6, 0, 0.8, 0), (6, 1, 0.8, 0),
(7, 0, 0.8, 0), (7, 1, 0.8, 0), (8, 0, 0.8, 0), (7, 1, 0.7, 0.1)

f=**36%** (21 steps): (3, 1, 0.48, 0.16), (3, 0, 0.48, 0), (3, 1, 0.48, 0),
(4, 0, 0.48, 0), (3, 1, 0.36, 0.12)

f=**6%** (22 steps): (4, 1, 0.08, 0.02), (4, 0, 0.08, 0), (3, 1, 0.06, 0.02)

f=**4%** (22 steps): (4, 1, 0.8, 0.2), (0, 1, 0, 0.2), (1, 0, 0.2, 0), (1, 1, 0.2, 0),
(2, 0, 0.2, 0), (2, 1, 0.2, 0), (3, 0, 0.2, 0), (3, 1, 0.2, 0),
(4, 0, 0.2, 0), (4, 1, 0.2, 0), (5, 0, 0.2, 0), (4, 1, 0.16, 0.04)

f=**7% or 28%** (23 steps): (4, 1, 0.35, 0.0875), (4, 0, 0.35, 0), (4, 1, 0.35, 0),
(5, 0, 0.35, 0), (4, 1, 0.28, 0.07)

f=**9%** (23 steps): (3, 1, 0.45, 0.15), (3, 0, 0.45, 0), (3, 1, 0.45, 0),
(4, 0, 0.45, 0), (4, 1, 0.45, 0), (5, 0, 0.45, 0),
(4, 1, 0.36, 0.09)

f=**1%** (24 steps): (2, 0, 0.125, 0), (1, 1, 1/16, 1/16), (1, 0, 1/16, 0),
(1, 1, 1/16, 0), (1, 0, 1/16, 0), (2, 1, 1/16, 0),
(3, 0, 1/16, 0), (3, 1, 1/16, 0), (4, 0, 1/16, 0), (4, 1, 1/16, 0),
(5, 0, 1/16, 0), (4, 1, 0.05, 1/80), (4, 0, 0.05, 0),
(4, 1, 0.05, 0), (5, 0, 0.05, 0), (4, 1, 0.04, 0.01)

f=**21%** (25 steps): (4, 1, 0.28, 0.07), (4, 0, 0.28, 0), (3, 1, 0.21, 0.07)

f=**18%** (25 steps): (3, 1, 0.24, 0.08), (4, 0, 0.24, 0), (3, 1, 0.18, 0.06)

$f=$**56%** (25 steps): (4, 0, 0.64, 0), (4, 1, 0.64, 0), (5, 0, 0.64, 0), (5, 1, 0.64, 0), (6, 0, 0.64, 0), (6, 1, 0.64, 0), (7, 0, 0.64, 0), (7, 1, 0.64, 0), (8, 0, 0.64, 0), (7, 1, 0.56, 0.08)

$f=$**27%** (25 steps): (3, 1, 0.36, 0.12), (3, 0, 0.36, 0), (3, 1, 0.36, 0), (4, 0, 0.36, 0), (3, 1, 0.27, 0.09)

$f=$**3%** (26 steps): (4, 1, 0.04, 0.01), (4, 0, 0.04, 0), (3, 1, 0.03, 0.01)

$f=$**14%** (27 steps): (3, 1, 0.21, 0.07), (3, 0, 0.21, 0), (2, 1, 0.14, 0.07)

$f=$**49%** (29 steps): (7, 1, 0.56, 0.08), (7, 0, 0.56, 0), (7, 1, 0.56, 0), (8, 0, 0.56, 0), (7, 1, 0.49, 0.07)

$f=$**42%** (29 steps): (4, 0, 0.48, 0), (4, 1, 0.48, 0), (5, 0, 0.48, 0), (5, 1, 0.48, 0), (6, 0, 0.48, 0), (6, 1, 0.48, 0), (7, 0, 0.48, 0), (7, 1, 0.48, 0), (8, 0, 0.48, 0), (7, 1, 0.42, 0.06)

(d) **Capacities of 8 ml and 5 ml.** There are solutions for n = 1–18, 20–22, 24, 25, 27, 30, 32–36, 40, 42, 44, 45, 50–52, 56, 58, 60, 64–66, 68, 70, 72, 75, 80, 83, 85 and 88–900.

$f=$**40% or 60%** (5 steps): (0, 5, 0, 0), (5, 0, 0, 0), (5, 5, 0, 0), (5, 5, 0, 1), (8, 2, 0.6, 0.4)

$f=$**15% or 85%** (7 steps): (5, 5, 0, 0), (8, 2, 0, 0), (8, 2, 1, 0), (5, 5, 0.625, 0.375), (8, 2, 0.85, 0.15)

$f=$**25% or 75%** (7 steps): (8, 0, 0, 0), (3, 5, 0, 0), (3, 0, 0, 0), (0, 3, 0, 0), (8, 3, 0, 0), (8, 3, 1, 0), (6, 5, 0.75, 0.25)

$f=$**30% or 70%** (9 steps): (8, 3, 0, 0), (6, 5, 0, 0), (6, 5, 0, 1), (8, 3, 0.4, 0.6), (6, 5, 0.3, 0.7)

$f=$**16% or 24%** (9 steps): (8, 2, 0.6, 0.4) (0, 2, 0, 0.4), (8, 2, 0, 0.4), (5, 5, 0, 0.4), (8, 2, 0.24, 0.16)

$f=$**42% or 58%** (10 steps): (6, 5, 0.3, 0.7), (8, 3, 0.58, 0.42)

$f=$**6% or 9%** (11 steps): (8, 2, 0.85, 0.15), (0, 2, 0, 0.15), (8, 2, 0, 0.15), (5, 5, 0, 0.15), (8, 2, 0.09, 0.06)

$f=$**5%** (11 steps): (6, 5, 0.3, 0.7), (6, 0, 0.3, 0), (1, 5, 0.05, 0.25)

$f=$**10%** (11 steps): (8, 2, 0.24, 0.16), (5, 5, 0.15, 0.25), (8, 2, 0.3, 0.1)

$f=$**20% or 80%** (11 steps): (8, 2, 0, 0), (0, 2, 0, 0), (2, 0, 0, 0), (2, 5, 0, 0), (7, 0, 0, 0), (7, 5, 0, 0), (7, 5, 0, 1), (8, 4, 0.2, 8)

$f=$**34%** (12 steps): (0, 2, 0, 0.4), (0, 5, 0, 0.4), (5, 0, 0.4, 0), (5, 5, 0.4, 0), (8, 2, 0.4, 0), (5, 5, 0.25, 0.15), (8, 2, 0.34, 0.06)

$f=$**35%** (12 steps): (0, 2, 0, 0.4), (2, 0, 0.4, 0), (2, 5, 0.4, 0), (7, 0, 0.4, 0), (7, 5, 0.4, 0), (8, 4, 0.4, 0), (7, 5, 0.35, 0.05)

$f = 36\%$ (12 steps): (8, 3, 0.4, 0.6), (0, 3, 0, 0.6), (8, 3, 0, 0.6), (6, 5, 0, 0.6), (8, 3, 0.24, 0.36)

$f = 18\%$ (13 steps): (8, 3, 0.24, 0.36), (6, 5, 0.18, 0.42)

$f = 50\%$ (13 steps): (6, 5, 0, 0), (6, 0, 0, 0), (1, 5, 0, 0), (1, 0, 0, 0), (0, 1, 0, 0), (8, 1, 0, 0), (8, 1, 1, 0), (4, 5, 0.5, 0.5)

$f = 4\%$ (13 steps): (7, 5, 0.35, 0.05), (8, 4, 0.36, 0.04)

$f = 90\%$ (13 steps): (7, 5, 0, 0), (8, 4, 0, 0), (8, 4, 1, 0), (7, 5, 0.875, 0.125), (8, 4, 0.9, 0.1)

$f = 66\%$ (13 steps): (8, 4, 0.2, 8), (7, 5, 0.175, 0.825), (8, 4, 0.34, 0.66)

$f = 3\%$ (15 steps): (6, 5, 0.18, 0.42), (6, 0, 0.18, 0), (1, 5, 0.03, 0.15)

$f = 51\%$ (15 steps): (0, 3, 0, 0.6), (0, 5, 0, 0.6), (5, 0, 0.6, 0), (5, 5, 0.6, 0), (8, 2, 0.6, 0), (5, 5, 0.375, 0.225), (8, 2, 0.51, 0.09)

$f = 45\%$ **or** 55% (15 steps): (4, 5, 0.5, 0.5), (8, 1, 0.9, 0.1), (4, 5, 0.45, 0.55)

$f = 12\%$ (15 steps): (8, 2, 0.24, 0.16), (0, 2, 0, 0.16), (8, 2, 0, 0.16), (5, 5, 0, 0.16), (8, 2, 0.096, 0.064), (5, 50.06, 0.1), (8, 2, 0.12, 0.04)

$f = 17\%$ **or** 83% (15 steps): (8, 4, 0.9, 0.1), (7, 5, 63/80, 17/80), (8, 4, 0.83, 0.17)

$f = 64\%$ (15 steps): (8, 4, 0.2, 0.8), (0, 4, 0, 0.8), (8, 4, 0, 0.8), (7, 5, 0, 0.8), (8, 4, 0.16, 0.64)

$f = 11\%$ **or** 89% (16 steps): (6, 5, 0.75, 0.25), (6, 0, 0.75, 0), (1, 5, 0.125, 0.625), (1, 0, 0.125, 0), (0, 1, 0, 0.125), (8, 1, 0, 0.125), (4, 5, 0.125), (8, 1, 0.1, 0.025), (4, 5, 0.05, 0.075), (8, 1, 0.11, 0.015)

$f = 1\%$ (16 steps): (1, 5, 0.05, 0.25), (1, 0, 0.05, 0), (0, 1, 0, 0.05), (8, 1, 0, 0.05), (4, 5, 0, 0.05), (8, 1, 0.04, 0.01)

$f = 2\%$ (16 steps): (6, 0, 0, 0), (6, 0, 1, 0), (1, 5, 1/6, 5/6), (1, 0, 1/6, 0), (0, 1, 0, 1/6), (8, 1, 0, 1/6), (4, 5, 0, 1/6), (8, 1, 2/15, 1/30), (4, 5, 1/15, 0.1), (8, 1, 11/75, 0.02)

$f = 14\%$ (16 steps): (0, 2, 0, 0.16), (2, 0, 0.16, 0), (2, 5, 0.16, 0), (7, 0, 0.16, 0), (7, 5, 0.16, 0), (8, 4, 0.16, 0), (7, 5, 0.14, 0.02)

$f = 88\%$ (16 steps): (8, 1, 0, 0), (8, 1, 0, 1), (4, 5, 0, 1), (8, 1, 0.8, 0.2), (4, 5, 0.4, 0.6), (8, 1, 0.88, 0.12)

$f = 44\%$ **or** 56% (17 steps): (8, 1, 0.88, 0.12), (4, 5, 0.44, 0.56)

$f = 7\%$ (17 steps): (8, 1, 0.11, 0.015), (4, 5, 0.055, 0.07)

f=**8% or 72%** (17 steps): (0, 4, 0, 0.8), (4, 0, 0.8, 0), (4, 5, 0.8, 0), (8, 1, 0.8, 0), (4, 5, 0.4, 0.4), (8, 1, 0.72, 0.08)

f=**33%** (18 steps): (8, 1, 0.88, 0.12), (8, 0, 0.88, 0), (3, 5, 0.33, 0.55)

f=**13%** (18 steps): (0, 2, 0, 0.16), (0, 5, 0, 0.16), (5, 0, 0.16, 0), (5, 5, 0.16, 0), (8, 2, 0.16, 0), (5, 5, 0.1, 0.06), (8, 2, 0.136, 0.024), (5, 5, 0.085, 0.075), (8, 2, 0.13, 0.03)

f=**68%** (18 steps): (0, 4, 0, 0.8), (0, 5, 0, 0.8), (5, 0, 0.8, 0), (5, 5, 0.8, 0), (8, 2, 0.8, 0), (5, 5, 0.5, 0.3), (8, 2, 0.68, 0.12)

f=**27%** (19 steps): (7, 5, 0, 0), (7, 5, 0, 1), (8, 4, 0.2, 0.8), (0, 4, 0, 0.8), (4, 0, 0.8, 0), (4, 5, 0.8, 0), (8, 1, 0.8, 0), (4, 5, 0.4, 0.4), (8, 1, 0.72, 0.08), (8, 0, 0.72, 0), (3, 5, 0.27, 0.45)

f=**32%** (20 steps): (4, 5, 0.4, 0.6), (4, 0, 0.4, 0), (0, 4, 0, 0.4), (8, 4, 0, 0.4), (7, 5, 0, 0.4), (8, 4, 0.08, 0.32)

f=**65%** (20 steps): (8, 2, 0.68, 0.12), (5, 5, 0.425, 0.375), (8, 2, 0.65, 0.15)

f=**22%** (21 steps): (4, 0, 0.4, 0), (4, 5, 0.4, 0), (8, 1, 0.4, 0), (4, 5, 0.2, 0.2), (8, 1, 0.36, 0.04), (4, 5, 0.18, 0.22)

f=**21%** (21 steps): (4, 5, 0.44, 0.56), (0, 5, 0, 0.56), (5, 0, 0.56, 0), (8, 0, 0.56, 0), (3, 5, 0.21, 0.35)

f=**52%** (24 steps): (8, 4, 0.16, 0.64), (0, 4, 0, 0.64), (0, 5, 0, 0.64), (5, 0, 0.64, 0), (5, 5, 0.64, 0), (8, 2, 0.64, 0), (5, 5, 0.4, 0.24), (8, 2, 0.544, 0.096), (5, 5, 0.34, 0.3), (8, 2, 0.52, 0.12)

(e) **Capacities of 12 ml and 5 ml**. There are solutions for n=1–10, 12–21, 24–28, 30, 32–36, 38–40, 45, 48–52, 56, 57, 60, 61, 64–68, 70, 74–76, 80 and 85.

f=**50%** (6 steps): (0, 5, 0, 0), (5, 0, 0, 0), (5, 5, 0, 0), (10, 0, 0, 0), (10, 0, 1, 0), (5, 5, 0.5, 0.5)

f=**40% or 60%** (7 steps): (10, 0, 0, 0), (10, 5, 0, 0), (10, 5, 0, 1), (12, 3, 0.4, 0.6)

f=**10% or 90%** (9 steps): (10, 5, 0, 0), (12, 3, 0, 0), (12, 3, 1, 0), (10, 5, 5/6, 1/6), (12, 3, 0.9, 0.1)

f=**25% or 75%** (9 steps): (12, 0, 0, 0), (7, 5, 0, 0), (7, 0, 0, 0), (2, 5, 0, 0), (2, 0, 0, 0), (0, 2, 0, 0), (12, 2, 0, 0), (12, 2, 1, 0), (9, 50.75, 0.25)

$f = $ **15% or 85%** (11 steps): (12, 3, 0.9, 0.1), (10, 5, 0.75, 0.25),
(12, 3, 0.85, 0.15)

$f = $ **45% or 55%** (11 steps): (12, 2, 0, 0), (9, 5, 0, 0), (9, 5, 0, 1),
(12, 2, 0.6, 0.4), (9, 5, 0.45, 0.55)

$f = $ **24% or 36%** (11 steps): (12, 3, 0.4, 0.6), (0, 3, 0, 0.6), (12, 3, 0, 0.6),
(10, 5, 0, 0.6), (12, 3, 0.24, 0.36)

$f = $ **35%** (11 steps): (0, 3, 0, 0.6), (3, 0, 0.6, 0), (12, 0, 0.6, 0),
(7, 5, 0.25, 0.35)

$f = $ **30% or 70%** (11 steps): (12, 3, 0.4, 0.6), (10, 5, 1/3, 2/3),
(12, 3, 0.6, 0.4), (10, 5, 0.5, 0.5), (12, 3, 0.7, 0.3)

$f = $ **13% or 87%** (12 steps): (9, 5, 0.75, 0.25), (12, 2, 0.9, 0.1),
(9, 5, 0.675, 0.325), (12, 2, 0.87, 0.13)

$f = $ **22% or 78%** (12 steps): (9, 5, 0.45, 0.55), (12, 2, 0.78, 0.22)

$f = $ **20%** (12 steps): (12, 3, 0.24, 0.36), (10, 5, 0.2, 0.4)

$f = $ **4% or 6%** (13 steps): (12, 3, 0.9, 0.1), (0, 3, 0, 0.1), (12, 3, 0, 0.1),
(10, 5, 0, 0.1), (12, 3, 0.04, 0.06)

$f = $ **5%** (13 steps): (10, 0, 0, 0), (10, 0, 1, 0), (5, 5, 0.5, 0.5), (5, 0, 0.5, 0),
(5, 5, 0.5, 0), (10, 0, 0.5, 0), (10, 5, 0.5, 0), (12, 3, 0.5, 0),
(10, 5, 5/12, 1/12), (12, 3, 0.45, 0.05)

$f = $ **14%** (13 steps): (12, 3, 0.24, 0.36), (12, 0, 0.24, 0), (7, 5, 0.14, 0.1)

$f = $ **16%** (13 steps): (12, 3, 0.6, 0.4), (0, 3, 0, 0.4), (12, 3, 0, 0.4),
(10, 5, 0, 0.4), (12, 3, 0.16, 0.24)

$f = $ **80%** (13 steps): (12, 3, 0, 0), (0, 3, 0, 0), (3, 0, 0, 0), (3, 5, 0, 0),
(8, 0, 0, 0), (8, 5, 0, 0), (8, 5, 0, 1), (12, 1, 0.8, 0.2)

$f = $ **3%** (15 steps): (12, 0, 0, 0), (12, 0, 1, 0), (7, 5, 7/12, 5/12), (7, 0, 7/12, 0),
(2, 5, 1/6, 5/12), (2, 0, 1/6, 0), (0, 2, 0, 1/6), (12, 2, 0, 1/6),
(9, 5, 0, 1/6), (12, 2, 0.1, 1/15), (0, 2, 0, 1/15),
(12, 2, 0, 1/15), (9, 5, 0, 1/15), (12, 2, 0.04, 2/75),
(9, 5, 0.03, 11/300)

$f = $ **12% or 18%** (15 steps): (12, 3, 0.7, 0.3), (0, 3, 0, 0.3), (12, 3, 0, 0.3),
(10, 5, 0, 0.3), (12, 3, 0.12, 0.18)

$f = $ **9%** (15 steps): (12, 3, 0.85, 0.15), (0, 3, 0, 0.15), (12, 3, 0, 0.15),
(10, 5, 0, 0.15), (12, 3, 0.06, 0.09)

$f = $ **26%** (15 steps): (12, 2, 0.78, 0.22), (9, 5, 0.585, 0.415), (9, 0, 0.585, 0),
(4, 5, 0.26, 0.325)

$f=$**29%** (15 steps): (12, 2, 0.87, 0.13), (9, 5, 261/400, 139/400), (9, 0, 261/400, 0), (4, 5, 0.29, 29/80)

$f=$**21%** (15 steps): (12, 3, 0.24, 0.36), (0, 3, 0, 0.36), (3, 0, 0.36, 0), (12, 0, 0.36, 0), (7, 5, 0.21, 0.15)

$f=$**42%** (15 steps): (10, 5, 0.2, 0.4), (12, 3, 0.36, 24), (10, 5, 0.3, 0.3), (12, 3, 0.42, 0.18)

$f=$**56%** (15 steps): (3, 0, 0.6, 0), (3, 5, 0.6, 0), (8, 0, 0.6, 0), (8, 5, 0.6, 0), (12, 1, 0.6, 0), (8, 5, 0.4, 0.2), (12, 1, 0.56, 0.04)

$f=$**54%** (16 steps): (0, 3, 0, 0.6), (0, 5, 0, 0.6), (5, 0, 0.6, 0), (5, 5, 0.6, 0), (10, 0, 0.6, 0), (10, 5, 0.6, 0), (12, 3, 0.6, 0), (10, 5, 0.5, 0.1), (12, 3, 0.54, 0.06)

$f=$**2%** (17 steps): (12, 3, 0.45, 0.05), (0, 3, 0, 0.05), (12, 3, 0, 0.05), (10, 5, 0, 0.05), (12, 3, 0.02, 0.03)

$f=$**7%** (17 steps): (12, 3, 0.04, 0.06), (10, 5, 1/30, 1/15), (12, 3, 0.06, 0.04), (10, 5, 0.05, 0.05), (12, 3, 0.07, 0.03)

$f=$**8%** (17 steps): (12, 2, 0.6, 0.4), (0, 2, 0, 0.4), (12, 2, 0, 0.4), (9, 5, 0, 0.4), (12, 2, 0.24, 0.16), (9, 5, 0.18, 0.22), (9, 0, 0.18, 0), (4, 5, 0.08, 0.1)

$f=$**28%** (17 steps): (12, 3, 0.16, 0.24), (10, 5, 2/15, 4/15) (12, 3, 0.24, 0.16) (10, 5, 0.2, 0.2), (12, 3, 0.28, 0.12)

$f=$**1%** (18 steps): (0, 3, 0, 0.1), (0, 5, 0, 0.1), (5, 0, 0.1, 0), (5, 5, 0.1, 0), (10, 0, 0.1, 0), (10, 5, 0.1, 0), (12, 3, 0.1, 0), (10, 5, 1/12, 1/60), (12, 3, 0.09, 0.01)

$f=$**11%** (18 steps): (12, 2, 0.78, 0.22), (0, 2, 0, 0.22), (0, 5, 0, 0.22), (5, 0, 0.22, 0), (5, 5, 0.22, 0), (10, 0, 0.22, 0), (5, 5, 0.11, 0.11)

$f=$**51%** (18 steps): (12, 3, 0.54, 0.06), (10, 5, 0.45, 0.15), (12, 3, 0.51, 0.09)

$f=$**17%** (19 steps): (10, 5, 0.2, 0.4), (10, 0, 0.2, 0), (10, 5, 0.2, 0), (12, 3, 0.2, 0), (10, 5, 1/6, 1/30), (12, 3, 0.18, 0.02), (10, 5, 0.15, 0.05), (12, 3, 0.17, 0.03)

$f=$**27%** (19 steps): (10, 5, 0.3, 0.3), (10, 0, 0.3, 0), (10, 5, 0.3, 0), (12, 3, 0.3, 0), (10, 5, 0.25, 0.05), (12, 3, 0.27, 0.03)

$f=$**64%** (20 steps): (9, 5, 0, 0), (9, 0, 0, 0), (4, 5, 0, 0), (4, 0, 0, 0), (0, 4, 0, 0), (12, 4, 0, 0), (11, 5, 0, 0), (11, 5, 0, 1), (12, 4, 0.2, 0.8), (0, 4, 0, 0.8), (12, 4, 0, 0.8), (11, 5, 0, 0.8), (12, 4, 0.16, 0.64)

f=**34%** (20 steps): (0, 3, 0, 0.4), (0, 5, 0, 0.4), (5, 0, 0.4, 0),
(5, 5, 0.4, 0), (10, 0, 0.4, 0), (10, 5, 0.4, 0),
(12, 3, 0.4, 0), (10, 5, 1/3, 1/15), (12, 3, 0.36, 0.04),
(10, 5, 0.3, 0.1), (12, 3, 0.34, 0.06)

f=**31%** (21 steps): (12, 3, 0.28, 0.12), (10, 5, 7/30, 1/6), (12, 3, 0.3, 0.1),
(10, 5, 0.25, 0.15), (12, 3, 0.31, 0.09)

f=**33%** (22 steps): (12, 3, 0.34, 0.06), (10, 5, 17/60, 7/60), (12, 3, 0.33, 0.07)

f=**32%** (24 steps): (9, 0, 0, 0), (9, 0, 1, 0), (4, 5, 4/9, 5/9), (4, 0, 4/9, 0),
(0, 4, 0, 4/9), (12, 4, 0, 4/9), (11, 5, 0, 4/9),
(12, 4, 4/45, 16/45), (0, 4, 0, 16/45), (4, 0, 16/45, 0),
(4, 5, 16/45, 0), (9, 0, 16/45, 0), (9, 5, 16/45, 0),
(12, 2, 16/45, 0), (9, 5, 4/15, 4/45), (12, 2, 0.32, 8/225)

f=**72%** (24 steps): (0, 4, 0, 0.8), (4, 0, 0.8, 0), (4, 5, 0.8, 0), (9, 0, 0.8, 0),
(9, 5, 0.8, 0), (12, 2, 0.8, 0), (9, 5, 0.6, 0.2),
(12, 2, 0.72, 0.08)

f=**49%** (25 steps): (12, 4, 0.2, 0.8), (11, 5, 11/60, 49/60),
(12, 4, 26/75, 49/75), (0, 4, 0, 49/75), (4, 0, 49/75, 0),
(4, 5, 49/75, 0), (9, 0, 49/75, 0), (9, 5, 49/75, 0),
(12, 2, 49/75, 0), (9, 5, 0.49, 49/300)

f=**48%** (27 steps): (0, 4, 0, 0.64), (4, 0, 0.64, 0), (4, 5, 0.64, 0),
(9, 0, 0.64, 0), (9, 5, 0.64, 0), (12, 2, 0.64, 0),
(9, 5, 0.48, 0.16)

f=**68%** (27 steps): (0, 4, 0, 0.8), (0, 5, 0, 0.8), (5, 5, 0.8, 0), (10, 0, 0.8, 0),
(10, 5, 0.8, 0), (12, 3, 0.8, 0), (10, 5, 2/3, 2/15),
(12, 3, 0.72, 0.08), (10, 5, 0.6, 0.2), (12, 3, 0.68, 0.12)

Chapter 7: Physics Puzzles

48
Boating Surprise

The rock either floats or is neutrally buoyant causing no rise in the water level of the swimming pool.

49
Balance Problems

Solutions are given in the table below.

	A	B	C	D	E	F
(a)	14	4	105	63	1	–
(b)	2	3	40	20	70	15
(c)	35	30	520	528	792	3
(d)	22	77	18	15	72	27
(e)	25	30	119	136	4	1
(f)	30	5	70	7	3	24
(g)	35	42	33	118	188	189
(h)	5	6	22	33	14	8
(i)	40	48	22	77	36	63
(j)	12	16	29	30	25	–
(k)	11	1	46	92	91	51
(l)	71	1	21	3	785	559
(m)	17	3	1	9	234	114
(n)	35	9	38	2	8	20

50
The Hanging Rod

(a) Generalize so that $AB=a$, $BC=b$, $CD=c$ and $AD=d$.
The horizontal distances give (1) $d-a \cos A+b \cos(A+B)=c \cos D$.
The vertical distances give (2) $b \cos(A+B) - a \cos A = c \sin D$.
The distance of M below the ceiling is $MN=h=a \sin A - (b/2) \sin(A+B)$.
The minimum energy position implies h is maximized: $\partial h/\partial A=0$.
Take $\partial/\partial A$ of (1) and (2) and eliminate $\partial D/\partial A$ to get $1+\partial B/\partial A=$
$a \sin(A+D)/(b \sin(A+B+D))$ and $2 \cos A \sin(A+B+D)=$
$\cos(A+B) \sin(A+D)$.
Manipulate this equation and use $C=2\pi-A-B-D$ to produce
$\cos A \sin C=\sin B \cos D$.

From triangle *BMP* we use the sine law to get $MP = b \sin B/(2 \cos A)$. From triangle *CMP* we use the sine law to get $MP = b \sin C/(2 \cos D)$. Equating these two gives the condition that the extension of *NM* goes through *P*: $\cos A \sin C = \sin B \cos D$, which was to be proved.

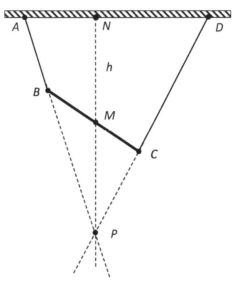

(b) Through manipulation of equations (1) and (2) we get the expression $(a^2 + b^2 + d^2 - c^2)/(2abd) - (1/b) \cos A = (1/a) \sin A \sin B - [(1/a) \cos A - 1/d)] \cos B$. This can be solved numerically to give $(A, B, C, D) = (72.496999°, 140.300406°, 85.1835968°, 62.0189978°)$.

51
Falling Ladders

(a) In Student *A*'s experiment we can follow θ and the coordinates of *C* as functions of time. Let the mass of the ladder be *m*, and the length of the ladder be *L* so its moment of inertia about *C* is $I = mL^2/12$. $C_x = (L/2) \cos \theta$ and $C_y = (L/2) \sin \theta$. Let $\omega = d\theta/dt$, $v_x = dC_x/dt$ and $v_y = dC_y/dt$. Since there are no horizontal forces on the ladder $v_x = 0$. Conservation of energy requires $mgL \sin \theta_0 = mgL \sin \theta + mv_y^2 + I\omega^2$. This leads to $\omega = - (4 \text{ g}/L)^{1/2} (\sin \theta_0 - \sin \theta)^{1/2} (1/3 + \cos^2 \theta)^{-1/2}$. The integral of $1/\omega$ with respect to θ from θ_0 to 0 gives the time for the ladder to fall. The time for the mass to fall is $T = (2L \sin \theta_0/\text{g})^{1/2}$. The integral must be done numerically for different values of θ_0 until it equals T. The value of θ_0 that makes the times equal is $\theta_0 = 70.8415°$.

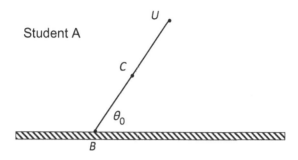

Student A

(b) In Student B's and Student C's experiments we use the same notation as in Student A's experiment. Conservation of energy requires in each case that $mgL \sin \theta_0 = mgL \sin \theta + m(v_x^2 + v_y^2) + I\omega^2$. This leads to $\omega = -(3 \, g/L)^{1/2}(\sin \theta_0 - \sin \theta)^{1/2}$. The integral of $1/\omega$ with respect to θ from θ_0 to 0 gives the time for the ladder to fall. Since this integral is the same for both experiments, the values of θ_0 determined by Students B and C must be the same. Again this integral must be done numerically. The value of θ_0 that makes the times equal is $\theta_0 = 47.9066°$.

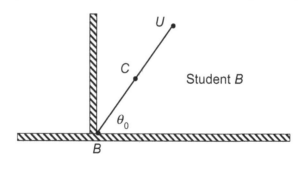

Student B

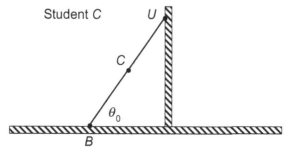

Student C

Chapter 8: Trapezoid Puzzles

52
Smallest IIT's

(a) The smallest IIT has $(r, b, a, c) = (1, 2, 1, 1)$.

(b) The next smallest IIT has $(r, b, a, c) = (13, 23, 1, 13)$ or $(13, 23, 1, 22)$.

(c) The next smallest IIT with $b = 2r$ has $(r, b, a, c) = (25, 50, 1, 35)$.

(d) The following set works.

$$r_{2n-1} = [(3+\sqrt{8})^{2n-1} + (3-\sqrt{8})^{2n-1} + 2]/8;$$
$$c_{2n-1} = [(3+\sqrt{8})^{2n-1} - (3-\sqrt{8})^{2n-1} + 2]/\sqrt{32}.$$

The first several entries are $(r, a, b, c) = (1, 2, 1, 1)$, $(25, 50, 1, 35)$, $(841, 1{,}682, 1, 1{,}189)$, $(28{,}561, 57{,}122, 1, 40{,}391)\ldots$

53
Smallest Circle

(a) $(r, b, a, c) = (1{,}012, 2{,}008, 1, 1{,}338)$ or $(1{,}012, 2{,}008, 1, 1{,}518)$.

(b) $(r, b, a, c) = (27{,}612, 51{,}128, 1, 30{,}798)$ or $(27{,}612, 51{,}128, 1, 45{,}838)$.

54
Prime IIT's

(a) $(r, b, a, c) = (7, 13, 2, 11)$ or $(7, 13, 11, 2)$.

(b) Unknown.

(c) Many cases such as $(r, b, a, c) = (7, 13, 2, 7)$, $(7, 13, 11, 7)$, $(13, 23, 1, 13)$ or $(19, 37, 11, 19)$.

(d) Unknown, but I could find none with $r < 20{,}000$.

55
Flat IIT's

(a) $(r, b, a, c) = (13, 23, 2, 1)$.

(b) The following set works.

$r_n = [(3 + 2\sqrt{3})(7 + 4\sqrt{3})^n - (3 - 2\sqrt{3})(7 - 4\sqrt{3})^n]/(4\sqrt{3})$;
$a_n = [(3 + 2\sqrt{3})(7 + 4\sqrt{3})^n + (3 - 2\sqrt{3})(7 - 4\sqrt{3})^n - 2]/4$; $b_n = a_n + 1$.

The first several entries are $(r, b, a, c) = (13, 24, 23, 1), (181, 314, 313, 1),$
$(2{,}521, 4{,}367, 4{,}366, 1)\ldots$

56
Pointed IIT's

(a) $(r, b, a, c) = (512, 141, 64, 1{,}016)$ having $c/b = 7.2$.

(b) $(r, b, a, c) = (3{,}523, 698, 157, 7{,}033)$ having $c/b = 10.08$.

(c) $r_n = n^3$, $b_n = 3n^2 - 1$, $a_n = n^2$ and $c_n = 2n^3 - n$ works for any n. c_n/b_n approaches $2n/3$ as n increases.

57
Nearly Square IIT's

(a) $(r, b, a, c, q) = (106, 149, 131, 159, 28/149 = 0.1879\ldots)$ and $(r, b, a, c, q) = (125, 182, 175, 175, 1/13 = 0.0769\ldots)$ are the two smallest such IIT's.

(b) There are several such series as shown below.

Consider $x_n^2 - 2y_n^2 = 1$. Then $x_n = [(3 + \sqrt{8})^n + (3 - \sqrt{8})^n]/2$ and $y_n = [(3 + \sqrt{8})^n - (3 - \sqrt{8})^n]/\sqrt{8}$.
Then $r_n = y_n^3$, $b_n = x_n y_n^2$, $a_n = x_n(y_n^2 - 1)$ and $c_n = b_n$ gives $q_n = 1/y_n^2$.
And $r'_n = x_n^3$, $b'_n = 2y_n(x_n^2 + 2)$, $a'_n = 2y_n x_n^2$ and $c'_n = a'_n$ gives $q'_n = 4/(x_n^2 + 2)$.
Also consider $x_n^2 - 2y_n^2 = -1$. Then $x_n = [(1 + \sqrt{2})^n + (1 - \sqrt{2})^n]/2$ and $y_n = [(1 + \sqrt{2})^n - (1 - \sqrt{2})^n]/\sqrt{8}$.
Then $r_n = y_n^3$, $b_n = x_n(y_n^2 + 1)$, $a_n = x_n y_n^2$ and $c_n = a_n$ gives $q_n = 2/(y_n^2 + 1)$.
Also $r'_n = x_n^3$, $b'_n = 2y_n x_n^2$, $a'_n = 2y_n(x_n^2 - 2)$ and $c'_n = b'_n$ gives $q'_n = 2/x_n^2$.

58
IIT's with Integer Altitudes

The figures below show the four smallest cases, all with $r=65$.
$(r,\ b,\ a,\ c)=(65,\ 126,\ 66,\ 50)$, $(65,\ 126,\ 66,\ 78)$, $(65,\ 112,\ 32,\ 50)$ or
$(65,\ 112,\ 32,\ 104)$,

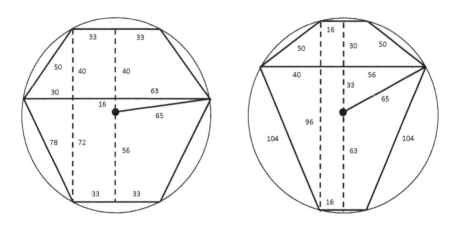

59
Smallest IIT's in a Square

(a) Type A: $(a, b, c, s)=(1, 2, 1, 2)$ or $(1, 2, 2, 2)$.

Type A

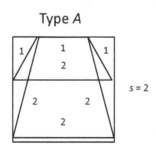

(b) Type B: $(a, b, c, s)=(1$ or $2, 11$ or $12, 13, 12)$.
    Type C: $(a, b, c, s)=(4, 10, 5, 8)$.
    Type D: $(a, b, c, s)=(7, 13, 17, 16)$.

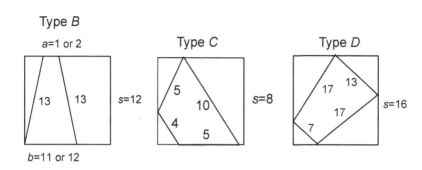

(c) Type A $(a, b, c, s) = (1, 11, 12, 11)$.

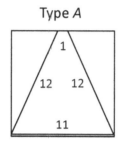

(d) Type A: $(a, b, c, s) = (1, 36, 40, 36)$.

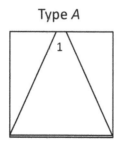

60
IIT's With $a=s$

(a) Type C: $(a, b, c, s)=(24, 30, 5, 24)$.

(b) Pick any $k>1$. Then $a_k=s_k=4k(4k^2-1)$, $b_k=4k(4k^2+1)$, $c_k=4k^2+1$. Another sequence that works for $k>1$ is $a_k=s_k=(2k+1)(4k^2+4k-3)$, $b_k=(2k+1)(4k^2+4k+5)$, $c_k=4k^2+4k+5$.

Type C

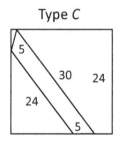

61
IIT's with $a = c$

(a) Type C: $(a, b, c, s)=(25, 55, 25, 44)$.

(b) Pick any m and n with $m > n$. Then $a_k=c_k=(m^2+n^2)^2$.

Case I. $b_k=(m^2+n^2)(3m^2-n^2)$ and $s_k=2mn(3m^2-n^2)$ if $m^2-n^2<2mn$.
Case II. $b_k=(m^2+n^2)(m^2+4mn+n^2)$ and $s_k=(m^2-n^2)(m^2+4mn+n^2)$ if $m^2-n^2>2mn$.

Type C

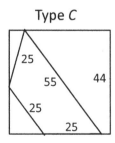

62
IIT's with x = u

(a) Type C: $(a, b, c, s) = (7, 25, 15, 20)$.

(b) Pick any m and n with $m > n$. Then $a_k = c_k = (m^2 + n^2)^2$.

Case I. $b_k = (m^2 + n^2)(3m^2 - n^2)$ and $s_k = 2mn(3m^2 - n^2)$ if $m^2 - n^2 < 2mn$.
Case II. $b_k = (m^2 + n^2)(m^2 + 4mn + n^2)$ and $s_k = (m^2 - n^2)(m^2 + 4mn + n^2)$ if $m^2 - n^2 > 2mn$.

Type C

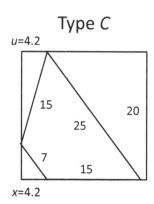

$u = 4.2$

15

20

25

7

15

$x = 4.2$

63
IIT's with b/s > 1.4

(a) Type C: $(a, b, c, s) = (17, 2{,}873, 2{,}028, 2{,}040)$. $b/s = 1.408333\ldots$

(b) Case I even k: $a_k = [(\sqrt{2} + 1)^{k+1} + (\sqrt{2} - 1)^{k+1})]/2$,
$b_k = [(\sqrt{2} + 1)^{3k+2} + (\sqrt{2} + 1)^k + (\sqrt{2} - 1)^k + (\sqrt{2} - 1)^{3k} + 2]/4$,
$c_k = [(\sqrt{2} + 1)^{3k+2} - (\sqrt{2} + 1)^k + (\sqrt{2} - 1)^k - (\sqrt{2} - 1)^{3k+2}]/(4\sqrt{2})$,
$s_k = [(\sqrt{2} + 1)^{3k+2} + (\sqrt{2} + 1)^{k+2} - (\sqrt{2} - 1)^{k+2} - (\sqrt{2} - 1)^{3k+2}]/(\sqrt{32})$.
Case II odd k: $a_k = [\sqrt{2} + 1)^{k+1} + (\sqrt{2} - 1)^{k+1})]/\sqrt{2}$,
$b_k = [(\sqrt{2} + 1)^{3k+2} + (\sqrt{2} + 1)^k + (\sqrt{2} - 1)^k + (\sqrt{2} - 1)^{3k+2}]/\sqrt{32}$,
$c_k = [(\sqrt{2} + 1)^{3k+2} - (\sqrt{2} + 1)^k + (\sqrt{2} - 1)^k - (\sqrt{2} - 1)^{3k+2}]/8$,
$s_k = [(\sqrt{2} + 1)^{3k+2} + (\sqrt{2} + 1)^{k+2} - (\sqrt{2} - 1)^{k+2} - (\sqrt{2} - 1)^{3k+2}]/8$.

Type C

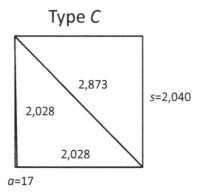

2,873

2,028

2,028

s=2,040

a=17

(c) Type D: For $k=3$ from the equations below, $(a, b, c, s) = (71, 16{,}969, 11{,}999, 12{,}049)$. $b/s = 1.4083326\ldots$

(d) Pick any $k > 1$. Then $a_k = [(\sqrt{2}+1)^{2k} - (\sqrt{2}-1)^{2k}) + \sqrt{8}\,]/\sqrt{8}$,

$b_k = [(\sqrt{2}+1)^{4k+1} + (\sqrt{2}-1)^{4k+1} + \sqrt{8}\,(\sqrt{2}+1)^{2k+1} - \sqrt{8}\,(\sqrt{2}-1)^{2k+1} - \sqrt{8}\,]/\sqrt{32}$,

$c_k = [(\sqrt{2}+1)^{4k+1} - (\sqrt{2}-1)^{4k+1} + \sqrt{8}\,(\sqrt{2}+1)^{2k+1} + \sqrt{8}\,(\sqrt{2}-1)^{2k+1} - 2]/8$,

$s_k = [(\sqrt{2}+1)^{4k+1} - (\sqrt{2}-1)^{4k+1} + (4\sqrt{2}-2)(\sqrt{2}+1)^{2k+1} + (4\sqrt{2}+2)(\sqrt{2}-1)^{2k+1} + 2]/8$.

Type D

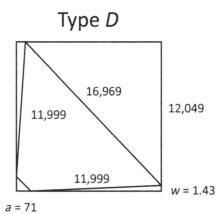

16,969

11,999

12,049

11,999

w = 1.43

a = 71

64
IIT's with $b/s < 0.8$, 0.75 and 0.71

Type *D*: For any $m > 2$ let $a_k = 2m^2 - 4m + 1$, $b_k = 2m^2 - 1$, $c_k = 2m^2 - 2m + 1$ and $s_k = \text{int}[2m(m-1)\sqrt{2}]$. This sequence has b/s decreasing and approaching $1/\sqrt{2}$ as *m* increases.

(a) For $m = 9$ we get $(a, b, c, s) = (127, 161, 145, 203)$ and $b/s = 0.7931034\ldots$.

(b) For $m = 18$ we get $(a, b, c, s) = (577, 647, 613, 865)$ and $b/s = 0.7479768\ldots$.

(c) For $m = 246$ we get $(a, b, c, s) = (120{,}049, 121{,}031, 120{,}541, 170{,}469)$ and $b/s = 0.7099883\ldots$.

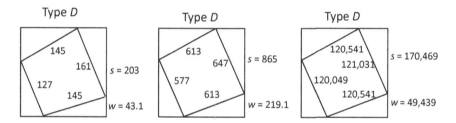

65
IIT's with Minimum Coverage

(a) Type *D*: For $k = 4$ in the treatment below, $(a, b, c, s) = (409, 569{,}737, 402{,}845, 403{,}154)$. Area$/s^2 = 0.500000255\ldots$.

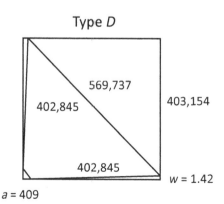

Type *D*

(b) Pick any $k > 1$. The resulting sequence given below produces solutions with trapezoids having successively lower fractions of the area of the square covered.

$a_k = [(\sqrt{2}+1)^{2k} - (\sqrt{2}-1)^{2k} + \sqrt{8}]/\sqrt{8}$,

$b_k = [(\sqrt{2}+1)^{4k+1} + (\sqrt{2}-1)^{4k+1} + \sqrt{8}\,(\sqrt{2}+1)^{2k+1} - \sqrt{8}\,(\sqrt{2}-1)^{2k+1} - \sqrt{8}]/\sqrt{32}$,

$c_k = [(\sqrt{2}+1)^{4k+1} - (\sqrt{2}-1)^{4k+1} + \sqrt{8}\,(\sqrt{2}+1)^{2k+1} + \sqrt{8}\,(\sqrt{2}-1)^{2k+1} - 2]/8$,

$s_k = [(\sqrt{2}+1)^{4k+1} - (\sqrt{2}-1)^{4k+1} + (\sqrt{32}-2)(\sqrt{2}+1)^{2k+1} + (\sqrt{32}+2)(\sqrt{2}-1)^{2k+1} + 2]/8$.

66
Largest Square

All Type D solutions can be generated from the set of Pythagorean Triangles as shown in the figure.

Let the triangle have sides e, h and c, $e = K(m^2 - n^2)$, $h = 2Kmn$ and $c = K(m^2 + n^2)$. Then from $s = (a+e)\sin\theta + h\cos\theta = (a+e)\cos\theta + h\sin\theta$ it follows that $a = h - e$ and $b = h + e$. From these it is possible to get a solution for any s from $s_{min} = \text{int}[h(h+e)/c+1]$ to $s_{max} = \text{int}(h\sqrt{2})$. The results below establish $s = 95$ as the largest integer-sided square that cannot circumscribe a Type D IIT.

$(m, n) = (7, 6)$ gives solutions for s from 96 to 118.
$(m, n) = (10, 1)$ gives solutions for s from 117 to 140.
$(m, n) = (8, 7)$ gives solutions for s from 126 to 158.
$(m, n) = (11, 2)$ gives solutions for s from 151 to 165.
$(m, n) = (9, 8)$ gives solutions for s from 160 to 203.
$(m, n) = (m, m-1)$ for all $m > 9$ gives solutions for s from 198 to ∞.
Examining all cases with $K = 1$ produces no possibilities with $s = 5$, 19 or 95.

Type *D*

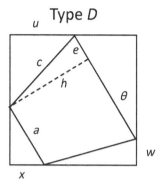

67
Multiple Solutions

The Type *D* trapezoid with $(a, b, c) = (7, 31, 25)$ will fit in integer sided squares with $s = 30, 31, 32$ or 33.

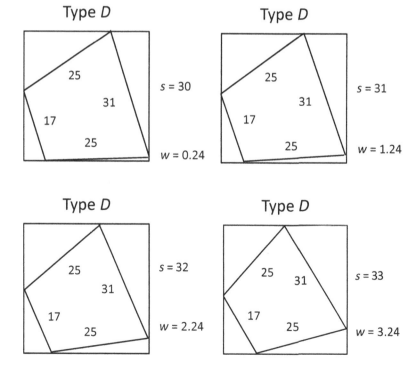

68
Smallest Squares

(a) Type *A*: The smallest two are $(a, b, c, s)=(1, 2, 1, 2)$ and $(1, 2, 2, 2)$ as shown below.

Type *A*

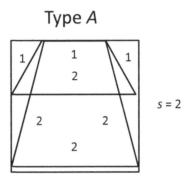

(b) As shown below there are 44 Type *A* solutions and one Type *C* solution for a square with $s=8$. The 44 Type *A* solutions come from varying *a* from 1 to 7 and putting *c* at *k* to 8 where *k* is the minimum *c* allowed for the chosen *a*.

Type *A*   Type *C*

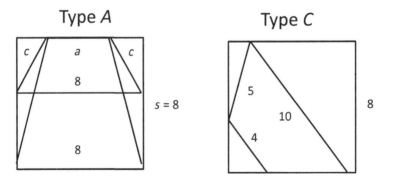

(c) As seen in the figure a Type A solution works with $(b-a)/a=0.01$.

Type *A*

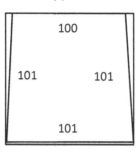

$s = 101$

(d) Type D: $(a, b, c, s) = (81{,}205, 81{,}607, 81{,}205, 81{,}606)$. $(b-a)/a = 0.0099751\ldots$.

(e) Pick any $k > 3$. The resulting sequence given below produces solutions with Type D IIT's having successively lower values of $(b-a)/a$.
$a_k = 2k^2 - 4k + 1$, $b_k = 2k^2 - 1$, $c_k = 2k^2 - 2k + 1$, $s_k = 2k^2 - 2$,
$(b_k - a_k)/a_k = 2/k + (3k-1)/(2k^3 - 4k^2 + k)$.

Type *D*

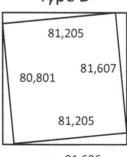

$s = 81{,}606$

(f) Type C: $(a, b, c, s) = (14, 20, 5, 16)$, $(a, b, c, s) = (8, 20, 10, 16)$.
Type D: $(a, b, c, s) = (8, 17, 13, 16)$.

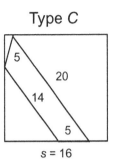

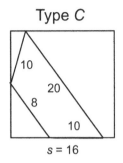

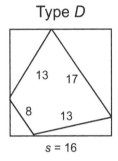

(g) Type *D*: $(a, b, c, s) = (17, 31, 25, 31)$.

(h) Type *D*: $(a, b, c, s) = (73, 161, 125, 161)$, $(a, b, c, s) = (127, 161, 145, 161)$.

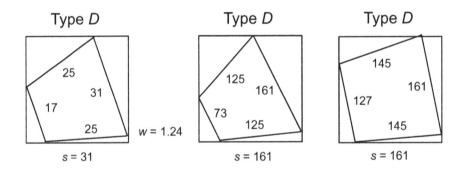

(i) Type *D*: $(a, b, c, s) = (7, 23, 17, 21)$, Type *D*: $(a, b, c, s) = (17, 31, 25, 31)$.

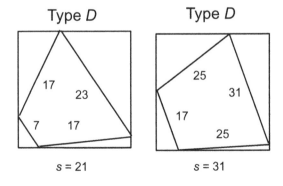

Chapter 9: Jeeps in the Desert

69
One-Way Trip with a Single Jeep

(a) Distance is 4/3. Go to P_1 at $x=1/3$, cache 1/3 units of fuel and return to A to refuel. Go to P_1, top off and go to $x=4/3$.

(b) Distance is 1.3. Go to P_1 at $x=0.3$, cache 0.3 units of fuel and return to A to refuel. Go to P_1, top off and go to $x=1.3$.

(c) 1.99 units of fuel. Go to P_1 at $x=0.33$, cache 0.33 units of fuel and return to A to refuel. Go to P_1, top off and go to $x=1.33$.

(d) Distance is 23/15. Go to P_1 at $x=0.2$, cache 0.6 units of fuel and return to A to refuel. Go to P_1, top off (leaving 0.4 units of fuel at P_1), go to P_2 at $x=1/5+1/3=8/15$, cache 1/3 units of fuel at P_2 and return to A (picking up 0.2 units of fuel at P_1 and leaving 0.2 units there). Refuel at A, go to P_1 to top off, go to P_2 to top off and go to point B at $x=23/15$.

(e) Distance is 43/30. Go to P_1 at $x=0.1$, cache 0.3 units of fuel and return to A to refuel. Go to P_1, top off (leaving 0.2 units of fuel at P_1), go to P_2 at $x=0.1+1/3=13/30$, cache 1/3 units of fuel at P_2 and return to A (picking up 0.1 units of fuel at P_1 and leaving 0.1 units there). Refuel at A, go to P_1 to top off, go to P_2 to top off and go to point B at $x=43/30$.

(f) 17/6 units of fuel. Go to P_1 at $x=1/6$, cache 0.5 units of fuel and return to A to refuel. Go to P_1, top off (leaving 1/3 units of fuel at P_1), go to P_2 at $x=1/6+1/3=0.5$, cache 1/3 units of fuel at P_2 and return to A (picking up 1/6 units of fuel at P_1 and leaving 1/6 units) Refuel at A, go to P_1 to top off, go to P_2 to top off and go to point B at $x=1.5$.

(g) $7+2{,}021/3{,}003$ units of fuel. This is best demonstrated by working backwards from B at $x=2$. This will require 7 cache points. To get to B we need 1 unit of fuel and the jeep at P_7 at $x=1$. To get that 1 unit at P_7 we need 2 units and the jeep at P_6 at $x=1-1/3$. To get 2 units at P_6 we need 3 units and the jeep at P_5 at $x=1-1/3-1/5$. This continues until we need 7 units and the jeep at P_1 at $x=1-1/3-1/5-1/7-1/9-1/11-1/13=2{,}021/45{,}045$. To accomplish this we first need 7 round trips from A to P_1 to A (each depositing 41,003/45,045 units) and a final trip to P_1 starting with a load of 2,021/3,003 units of fuel.

70
Round Trip with a Single Jeep

(a) Distance is 3/4. Go to P_1 at $x=1/4$, cache 1/2 units of fuel and return to A to refuel. Go to P_1, top off (leaving 1/4 units), go to $x=3/4$ and return to A picking up fuel at P_1.

(b) Distance is 29/40. Go to P_1 at $x=9/40$, cache 9/20 units of fuel and return to A to refuel. Go to P_1, top off (leaving 9/40 units), go to $x=29/40$ and return to A picking up fuel at P_1.

(c) 1.8 units of fuel. Go to P_1 at $x=0.2$, cache 0.4 units of fuel and return to A to refuel. Go to P_1, top off (leaving 0.2 units), go to B at $x=0.7$ and return to A picking up fuel at P_1.

(d) Distance is 5/6. Go to P_1 at $x=1/12$, cache 1/3 units of fuel and return to A to refuel. Go to P_1, leave another 5/6 units of fuel at P_1 for a total of 7/6 units and return to A to refuel. Go to P_1, top off (leaving 13/12 units of fuel at P_1), go to P_2 at $x=1/12+1/4$, cache 1/2 units of fuel at P_2 and return to P_1. Top off at P_1, (leaving 1/12 units of fuel at P_1), top off at P_2 (leaving 1/4 units of fuel), go to B at $x=1/12+1/4+1/2=5/6$ and return to A.

(e) 11/3 units of fuel. This is demonstrated by working backwards from B at $x=1$. This will require 3 cache points (at 1/12, 1/12+1/6 and 1/12+1/6+1/4). To get to B we need 5/4 units of fuel and the jeep at P_3 at $x=1/2$. To get 5/4 units at P_3 we need 13/6 units and the jeep at P_2 at $x=1/2-1/4=1/4$. To get 13/6 units at P_2 we need 37/12 units and the jeep at P_1 at $x=1/2-1/4-1/6=1/12$. To get 37/12 units at P_1 we first need 3 round trips from A to P_1 to A (depositing 5/6, 5/6 and 5/6 units) and a final trip to P_1 starting with a load of 2/3 units of fuel. Total fuel needed is 37/12+7/12=11/3.

(f) 10+1,969/2,520 units of fuel. This is demonstrated by working backwards from B at $x=3/2$. This will require 10 cache points (P_{10} at

$x=1$, P_9 at $x=1-1/4$, P_8 at $x=1-1/4-1/6$, and so on to $P_1=1-1/4-1/6-1/8-1/10\ldots-1/20=179/5{,}040$.) To get to B we need $1+1/4$ units of fuel and the jeep at P_{10} at $x=1/2$. To get that $1+1/4$ units at P_{10} we need $2+1/6$ units at P_9 and so on until we need $10+179/5{,}040$ units of fuel at P_1. To get this fuel at P_1 make 10 round trips from A to P_1 to A, (each depositing $2341/2520$ units of fuel) and a final trip from A to P_1 starting with a load of $3938/5040$ units. Total fuel needed is $10+179/5{,}040+21\times179/5{,}040=10+1{,}969/2{,}520$.

71
A Single One-Way Trip with Two Jeeps

(a) Distance is 4/3. Jeep 1 goes to P_1 at $x=1/3$ and stays there. Jeep 2 makes a round trip from A to P_1 to A, topping off Jeep 1 at P_1. Then Jeep 1 goes to B at a distance of 4/3.

(b) Distance is 25/18. Jeep 1 goes to P_1 at $x=1/6$ and stays there. Jeep 2 makes a round trip from A to P_1 to A, topping off Jeep 1 at P_1. Then Jeep 1 goes to P_2 at $x=1/6+2/9=7/18$. Jeep 2 makes a round trip from A to P_2 to A, topping off Jeep 1 at P_2. Then Jeep 1 goes to B at a distance of 25/18.

(c) 3 units of fuel. Jeep 1 goes to P_1 at $x=1/3$ and stays there. Jeep 2 makes a round trip from A to P_1 to A, topping off Jeep 1 at P_1. Then Jeep 1 goes to P_2 at $x=1/3+1/9=4/9$. Jeep 2 makes a round trip from A to P_2 to A, topping off Jeep 1 at P_2. Then Jeep 1 goes to B at a distance of 13/9.

(d) 3.69 units of fuel. Jeep 1 goes to P_1 at $x=0.23$ and stays there. Jeep 2 makes a round trip from A to P_1 to A, topping off Jeep 1 at P_1. Then Jeep 1 goes to P_2 at $x=0.41$. Jeep 2 makes a round trip from A to P_2 to A, topping off Jeep 1 at P_2. Then Jeep 1 goes to P_3 at $x=0.47$. Jeep 2 makes a round trip from A to P_3 to A, topping off Jeep 1 at P_3. Then Jeep 1 goes to B at a distance of 1.47.

(e) 4.609 units of fuel. Jeep 1 goes to P_1 at $x=0.203$ and stays there. Jeep 2 makes a round trip from A to P_1 to A, topping off Jeep 1 at P_1. Then Jeep 1 goes to P_2 at $x=0.401$. Jeep 2 makes a round trip from A to P_2 to A, topping off Jeep 1 at P_2. Then Jeep 1 goes to P_3 at $x=0.467$. Jeep 2 makes a round trip from A to P_3 to A, topping off Jeep 1 at P_3. Then Jeep 1 goes to P_4 at $x=0.489$. Jeep 2 makes a round trip from A to P_4 to A, topping off Jeep 1 at P_4. Then Jeep 1 goes to B at a distance of 1.489. Fuel required is 4.609 tanks.

72
Two One-Way Trips with Two Jeeps

(a) Distance is 7/6. Jeep 1 goes to P_1 at $x=1/3$ and stays there. Jeep 2 makes a round trip from A to P_1 to A, topping off Jeep 1 at P_1. After refueling at A Jeep 2 goes to P_1. Total fuel at P_1 is 5/3 units which is split equally between the jeeps, allowing them to get to B at a distance from A of $1/3+5/6=7/6$. Fuel required is 3 tanks.

(b) 4 units of fuel. Jeep 1 goes to P_1 at $x=1/3$ and stays there. Jeep 2 makes a round trip from A to P_1 to A, topping off Jeep 1 at P_1. Then Jeep 1 goes to P_2 at $x=4/9$. Jeep 2 makes a round trip from A to P_2 to A, topping off Jeep 1 at P_2. After refueling at A Jeep 2 goes to P_2. Total fuel at P_3 is 14/9 units which is split equally between the jeeps, allowing them to get to B at a distance from A of $4/9+7/9=11/9$. Fuel required is 4 tanks.

(c) 4.42 units of fuel. Jeep 1 goes to P_1 at $x=0.14$ and stays there. Jeep 2 makes a round trip from A to P_1 to A, topping off Jeep 1 at P_1. Then Jeep 1 goes to P_2 at $x=0.38$. Jeep 2 makes a round trip from A to P_2 to A, topping off Jeep 1 at P_2. Then Jeep 1 goes to P_3 at $x=0.46$. Jeep 2 makes a round trip from A to P_3 to A, topping off Jeep 1 at P_3. After refueling at A Jeep 2 goes to P_3. Total fuel at P_3 is 1.54 units which is split equally between the jeeps, allowing them to get to B at a distance from A of $0.46+0.77=1.23$. Fuel required is 4.42 tanks.

(d) Distance is 67/54. Jeep 1 goes to P_1 at $x=1/3$ and stays there. Jeep 2 makes a round trip from A to P_1 to A, topping off Jeep 1 at P_1. Then Jeep 1 goes to P_2 at $x=4/9$. Jeep 2 makes a round trip from A to P_2 to A, topping off Jeep 1 at P_2. Then Jeep 1 goes to P_3 at $x=13/27$. Jeep 2 makes a round trip from A to P_3 to A, topping off Jeep 1 at P_3. After refueling at A Jeep 2 goes to P_3. Total fuel at P_3 is 41/27 units which is split equally between the jeeps, allowing them to get to B at a distance from A of $13/27+41/54=67/54$. Fuel required is 5 tanks.

(e) Distance is 1,339/1,080. Jeep 1 goes to P_1 at $x=19/60$ and stays there. Jeep 2 makes a round trip from A to P_1 to A, topping off Jeep 1 at P_1. Then Jeep 1 goes to P_2 at $x=79/180$. Jeep 2 makes a round trip from A to P_2 to A, topping off Jeep 1 at P_2. Then Jeep 1 goes to P_3 at $x=259/540$. Jeep 2 makes a round trip from A to P_3 to A, topping off Jeep 1 at P_3. After refueling at A Jeep 2 goes to P_3. Total fuel at P_3 is 821/540 units which is split equally between the jeeps, allowing them to get to B at a distance from A of $259/540+821/1,080=1,339/1,080$. Fuel used is 4.95 tanks.

(f) 6.095 units of fuel. Jeep 1 goes to P_1 at $x=0.365$ and stays there. Jeep 2 makes a round trip from A to P_1 to A, topping off Jeep 1 at P_1. Then Jeep 1 goes to P_2 at $x=0.455$. Jeep 2 makes a round trip from A to P_2 to A, topping off Jeep 1 at P_2. Then Jeep 1 goes to P_3 at $x=0.485$. Jeep 2 makes a round trip from A to P_3 to A, topping off Jeep 1 at P_3. Then Jeep 1 goes to P_4 at $x=0.495$. Jeep 2 makes a round trip from A to P_4 to A, topping off Jeep 1 at P_4. After refueling at A Jeep 2 goes to P_4. Total fuel at P_4 is 1.505 units which is split equally between the jeeps, allowing them to get to B at a distance from A of $0.495+0.7525=1.2475$. Fuel required is 6.095 tanks.

73
One Round Trip with Two Jeeps

(a) Distance is 8/9. Jeep 1 goes to P_1 at $x=1/6$ and stays there. Jeep 2 makes a round trip from A to P_1 to A, topping off Jeep 1 at P_1. Then Jeep 1 goes to P_2 at $x=7/18$. Jeep 2 makes a round trip from A to P_2 to A, topping off Jeep 1 at P_2. Then Jeep 1 goes to B at a distance of 8/9 and returns to P_2. Jeep 2 makes a round trip from A to P_2 to A, giving Jeep 1 enough fuel to reach P_1. Jeep 1 then goes to P_1. Jeep 2 makes a round trip from A to P_1 to A, giving Jeep 1 enough fuel to reach A. Jeep 1 then goes to A. Fuel required is 3 tanks.

(b) Distance is 11/12. Jeep 1 goes to P_1 at $x=1/4$ and stays there. Jeep 2 makes a round trip from A to P_1 to A, topping off Jeep 1 at P_1. Then Jeep 1 goes to P_2 at $x=1/4+1/6=5/12$. Jeep 2 makes a round trip from A to P_2 to A, topping off Jeep 1 at P_2. Then Jeep 1 goes to B at a distance of 11/12 and returns to P_2. Jeep 2 makes a round trip from A to P_2 to A, giving Jeep 1 enough fuel to reach P_1. Jeep 1 then goes to P_1. Jeep 2 makes a round trip from A to P_1 to A, giving Jeep 1 enough fuel to reach A. Jeep 1 then goes to A. Fuel required is 3.5 tanks.

(c) 5.11 units of fuel. Jeep 1 goes to P_1 at $x=0.185$ and stays there. Jeep 2 makes a round trip from A to P_1 to A, topping off Jeep 1 at P_1. Then Jeep 1 goes to P_2 at $x=0.395$. Jeep 2 makes a round trip from A to P_2 to A, topping off Jeep 1 at P_2. Then Jeep 1 goes to P_3 at $x=0.465$. Jeep 2 makes a round trip from A to P_3 to A, topping off Jeep 1 at P_3. Then Jeep 1 goes to B at a distance of 0.965 and returns to P_3. Jeep 2 makes a round trip from A to P_3 to A, giving Jeep 1 at P_3 enough fuel to reach P_2. Jeep 1 then goes to P_2. Jeep 2 makes a round trip from A to P_2 to A, giving Jeep 1 at P_2 enough fuel to reach P_1. Jeep 1 then goes to P_1. Jeep 2 makes a round trip from A to P_1 to A, giving Jeep 1 at P_1 enough fuel to reach A. Jeep 1 then goes to A. Fuel required is 5.11 tanks.

(d) 6 units of fuel. Jeep 1 goes to P_1 at $x=1/3$ and stays there. Jeep 2 makes a round trip from A to P_1 to A, topping off Jeep 1 at P_1. Then Jeep 1 goes to P_2 at $x=1/3+1/9=4/9$. Jeep 2 makes a round trip from A to P_2 to A, topping off Jeep 1 at P_2. Then Jeep 1 goes to P_3 at $x=4/9+1/27=13/27$. Jeep 2 makes a round trip from A to P_3 to A, topping off Jeep 1 at P_3. Then Jeep 1 goes to B at a distance of 53/54 and returns to P_3. Jeep 2 makes a round trip from A to P_3 to A, giving Jeep 1 at P_3 enough fuel to reach P_2. Jeep 1 then goes to P_2. Jeep 2 makes a round trip from A to P_2 to A, giving Jeep 1 at P_2 enough fuel to reach P_1. Jeep 1 then goes to P_1. Jeep 2 makes a round trip from A to P_1 to A, giving Jeep 1 at P_1 enough fuel to reach A. Jeep 1 then goes to A. Fuel required is 6 tanks.

(e) 6.3 units of fuel. Jeep 1 goes to P_1 at $x=0.05$ and stays there. Jeep 2 makes a round trip from A to P_1 to A, topping off Jeep 1 at P_1. Then Jeep 1 goes to P_2 at $x=0.35=13/30$. Jeep 2 makes a round trip from A to P_2 to A, topping off Jeep 1 at P_2. Then Jeep 1 goes to P_3 at $x=0.45$.

Jeep 2 makes a round trip from A to P_3 to A, topping off Jeep 1 at P_3. Then Jeep 1 goes to P_4 at $x=0.45+1/30=29/60$. Jeep 2 makes a round trip from A to P_4 to A, topping off Jeep 1 at P_4. Then Jeep 1 goes to B at a distance of 59/60 and returns to P_4. Jeep 2 makes a round trip from A to P_4 to A, giving Jeep 1 at P_4 enough fuel to reach P_3. Jeep 1 then goes to P_3. Jeep 2 makes a round trip from A to P_3 to A, giving Jeep 1 at P_3 enough fuel to reach P_2. Jeep 1 then goes to P_2. Jeep 2 makes a round trip from A to P_2 to A, giving Jeep 1 at P_2 enough fuel to reach P_1. Jeep 1 then goes to P_1. Jeep 2 makes a round trip from A to P_1 to A, giving Jeep 1 at P_1 enough fuel to reach A. Jeep 1 then goes to A. Fuel required is 6.3 tanks.

(f) Distance is 134/135. Jeep 1 goes to P_1 at $x=0.3$ and stays there. Jeep 2 makes a round trip from A to P_1 to A, topping off Jeep 1 at P_1. Then Jeep 1 goes to P_2 at $x=0.3+2/15=13/30$. Jeep 2 makes a round trip from A to P_2 to A, topping off Jeep 1 at P_2. Then Jeep 1 goes to P_3 at $x=13/30+2/45=43/90$. Jeep 2 makes a round trip from A to P_3 to A, topping off Jeep 1 at P_3. Then Jeep 1 goes to P_4 at $x=43/90+2/135=133/270$. Jeep 2 makes a round trip from A to P_4 to A, topping off Jeep 1 at P_4. Then Jeep 1 goes to B at a distance of 134/135 and returns to P_4. Jeep 2 makes a round trip from A to P_4 to A, giving Jeep 1 at P_4 enough fuel to reach P_3. Jeep 1 then goes to P_3. Jeep 2 makes a round trip from A to P_3 to A, giving Jeep 1 at P_3 enough fuel to reach P_2. Jeep 1 then goes to P_2. Jeep 2 makes a round trip from A to P_2 to A, giving Jeep 1 at P_2 enough fuel to reach P_1. Jeep 1 then goes to P_1. Jeep 2 makes a round trip from A to P_1 to A, giving Jeep 1 at P_1 enough fuel to reach A. Jeep 1 then goes to A. Fuel required is 7.8 tanks.

(g) 7.38 units of fuel. Jeep 1 goes to P_1 at $x=0.23$ and stays there. Jeep 2 makes a round trip from A to P_1 to A, topping off Jeep 1 at P_1. Then Jeep 1 goes to P_2 at $x=0.41$. Jeep 2 makes a round trip from A to P_2 to A, topping off Jeep 1 at P_2. Then Jeep 1 goes to P_3 at $x=0.47$. Jeep 2 makes a round trip from A to P_3 to A, topping off Jeep 1 at P_3. Then Jeep 1 goes to P_4 at $x=0.49$. Jeep 2 makes a round trip from A to P_4 to A, topping off Jeep 1 at P_4. Then Jeep 1 goes to B at a distance of 0.99 and returns to P_4. Jeep 2 makes a round trip from A to P_4 to A, giving Jeep 1 at P_4 enough fuel to reach P_3. Jeep 1 then goes to P_3. Jeep 2 makes a round trip from A to P_3 to A, giving Jeep 1 at P_3 enough fuel to reach P_2. Jeep 1 then goes to P_2. Jeep 2 makes a round trip from A to P_2 to A, giving Jeep 1 at P_2 enough fuel to reach P_1. Jeep 1 then goes to P_1. Jeep 2 makes a round trip from A to P_1 to A, giving Jeep 1 at P_1 enough fuel to reach A. Jeep 1 then goes to A. Fuel required is 7.38 tanks.

74
Two Round Trips with Two Jeeps

(a) Distance is 2/3. Jeep 1 goes to P_1 at $x=1/3$ and stays there. Jeep 2 makes a round trip from A to P_1 to A, topping off Jeep 1 at P_1. Jeep 2 then goes to P_1. Both jeeps go to B at a distance of 2/3 and return to P_1. Jeep 2 goes to A and makes a round trip from A to P_1 to A, giving Jeep 1 enough fuel to reach A. Jeep 1 then goes to A. Fuel required is 4 tanks.

(b) 5.2 units of fuel. Jeep 1 goes to P_1 at $x=0.2$ and stays there. Jeep 2 makes a round trip from A to P_1 to A, topping off Jeep 1 at P_1. Jeep 1 goes to P_2 at a distance of 0.4. Jeep 2 makes a round trip from A to P_2 to A, topping off Jeep 1 at P_2. Jeep 2 goes to P_2. Both jeeps go to B at a distance of 0.7 units and return to P_2. Jeep 2 goes to A and makes a round trip from A to P_2 to A, giving Jeep 1 enough fuel to reach P_1. Jeep 1 goes to P_1. Jeep 2 makes a round trip from A to P_1 to A, giving Jeep 1 enough fuel to reach A. Jeep 1 then goes to A. Fuel required is 5:2 tanks.

(c) Distance is 13/18. Jeep 1 goes to P_1 at $x=1/3$ and stays there. Jeep 2 makes a round trip from A to P_1 to A, topping off Jeep 1 at P_1. Jeep 1 goes to P_2 at a distance of 4/9. Jeep 2 makes a round trip from A to P_2 to A, topping off Jeep 1 at P_2. Jeep 2 goes to P_2. Both jeeps go to B at a distance of 13/18 units and return to P_2. Jeep 2 goes to A and makes a round trip from A to P_2 to A, giving Jeep 1 enough fuel to reach P_1. Jeep 1 goes to P_1. Jeep 2 makes a round trip from A to P_1 to A, giving Jeep 1 enough fuel to reach A. Jeep 1 then goes to A. Fuel required is 6 tanks.

(d) Distance is 79/108. Jeep 1 goes to P_1 at $x=1/6$ and stays there. Jeep 2 makes a round trip from A to P_1 to A, topping off Jeep 1 at P_1. Jeep 1 goes to P_2 at a distance of 7/18. Jeep 2 makes a round trip from A to P_2

to A, topping off Jeep 1 at P_2. Jeep 1 goes to P_3 at a distance of 25/54. Jeep 2 makes a round trip from A to P_3 to A, topping off Jeep 1 at P_3. Jeep 2 goes to P_3. Both jeeps go to B at a distance of 79/108 units and return to P_3. Jeep 2 goes to A and makes a round trip from A to P_3 to A, giving Jeep 1 enough fuel to reach P_2. Jeep 1 goes to P_2. Jeep 2 makes a round trip from A to P_2 to A, giving Jeep 1 enough fuel to reach P_1. Jeep 1 goes to P_1. Jeep 2 makes a round trip from A to P_1 to A, giving Jeep 1 enough fuel to reach A. Jeep 1 goes to A. Fuel required is 7 tanks.

(e) 7.92 units of fuel. Jeep 1 goes to P_1 at $x=0.32$ and stays there. Jeep 2 makes a round trip from A to P_1 to A, topping off Jeep 1 at P_1. Jeep 1 goes to P_2 at a distance of 0.44. Jeep 2 makes a round trip from A to P_2 to A, topping off Jeep 1 at P_2. Jeep 1 goes to P_3 at a distance of 0.48. Jeep 2 makes a round trip from A to P_3 to A, topping off Jeep 1 at P_3. Jeep 2 goes to P_3. Both jeeps go to B at a distance of 0.74 units and return to P_3. Jeep 2 goes to A and makes a round trip from A to P_3 to A, giving Jeep 1 enough fuel to reach P_2. Jeep 1 goes to P_2. Jeep 2 makes a round trip from A to P_2 to A, giving Jeep 1 enough fuel to reach P_1. Jeep 1 goes to P_1. Jeep 2 makes a round trip from A to P_1 to A, giving Jeep 1 enough fuel to reach A. Jeep 1 goes to A. Fuel required is 7.92 tanks.

(f) Distance is 67/90. Jeep 1 goes to P_1 at $x=0.2$ and stays there. Jeep 2 makes a round trip from A to P_1 to A, topping off Jeep 1 at P_1. Jeep 1 goes to P_2 at a distance of 0.4. Jeep 2 makes a round trip from A to P_2 to A, topping off Jeep 1 at P_2. Jeep 1 goes to P_3 at a distance of $0.4+1/15=7/15$. Jeep 2 makes a round trip from A to P_3 to A, topping off Jeep 1 at P_3. Jeep 1 goes to P_4 at a distance of $7/15+1/45=22/45$. Jeep 2 makes a round trip from A to P_4 to A, topping off Jeep 1 at P_4. Jeep 2 goes to P_4. Both jeeps go to B at a distance of 67/90 units and return to P_4. Jeep 2 goes to A and makes a round trip from A to P_4 to A, giving Jeep 1 enough fuel to reach P_3. Jeep 1 goes to P_3. Jeep 2 makes a round trip from A to P_3 to A, giving Jeep 1 enough fuel to reach P_2. Jeep 1 goes to P_2. Jeep 2 makes a round trip from A to P_2 to A, giving Jeep 1 enough fuel to reach P_1. Jeep 1 goes to P_1. Jeep 2 makes a round trip from A to P_1 to A, giving Jeep 1 enough fuel to reach A. Jeep 1 goes to A. Fuel required is 9.2 tanks.

(g) Distance is 403/540. Jeep 1 goes to P_1 at $x=0.3$ and stays there. Jeep 2 makes a round trip from A to P_1 to A, topping off Jeep 1 at P_1. Jeep 1 goes to P_2 at a distance of 13/30. Jeep 2 makes a round trip from A to P_2 to A, topping off Jeep 1 at P_2. Jeep 1 goes to P_3 at a distance of $13/30+2/45=43/90$. Jeep 2 makes a round trip from A to P_3 to A, topping off Jeep 1 at P_3. Jeep 1 goes to P_4 at a distance of $43/90+2/135=133/270$. Jeep 2 makes a round trip from A to P_4 to A, topping off Jeep 1 at P_4. Jeep 2 goes to P_4. Both jeeps go to B at a distance of 403/540 units and return to P_4. Jeep 2 goes to A and makes a round trip from A to P_4 to A, giving Jeep 1 enough fuel to reach P_3. Jeep 1 goes to P_3. Jeep 2 makes a round trip from A to P_3 to A, giving Jeep 1 enough fuel to reach P_2. Jeep 1 goes to P_2. Jeep 2 makes a round trip from A to P_2 to A, giving Jeep 1 enough fuel to reach P_1. Jeep 1 goes to P_1. Jeep 2 makes a round trip from A to P_1 to A, giving Jeep 1 enough fuel to reach A. Jeep 1 goes to A. Fuel required is 9.8 tanks.

(h) 10 units. Jeep 1 goes to P_1 at $x=1/3$ and stays there. Jeep 2 makes a round trip from A to P_1 to A, topping off Jeep 1 at P_1. Jeep 1 goes to P_2 at a distance of 4/9. Jeep 2 makes a round trip from A to P_2 to A, topping off Jeep 1 at P_2. Jeep 1 goes to P_3 at a distance of $4/9+1/27=13/27$. Jeep 2 makes a round trip from A to P_3 to A, topping off Jeep 1 at P_3. Jeep 1 goes to P_4 at a distance of $13/27+1/81=40/81$. Jeep 2 makes a round trip from A to P_4 to A, topping off Jeep 1 at P_4. Jeep 2 goes to P_4. Both jeeps go to B at a distance of 121/162 units and return to P_4. Jeep 2 goes to A and makes a round trip from A to P_4 to A, giving Jeep 1 enough fuel to reach P_3. Jeep 1 goes to P_3. Jeep 2 makes a round trip from A to P_3 to A, giving Jeep 1 enough fuel to reach P_2. Jeep 1 goes to P_2. Jeep 2 makes a round trip from A to P_2 to A, giving Jeep 1 enough fuel to reach P_1. Jeep 1 goes to P_1. Jeep 2 makes a round trip from A to P_1 to A, giving Jeep 1 enough fuel to reach A. Jeep 1 goes to A. Fuel required is 10 tanks.

75
Two Jeeps and Two Depots

(a) 3.5 units. Jeep 1 goes to P_1 at $x=1/4$ and stays there. Jeep 2 makes a round trip from A to P_1 to A, topping off Jeep 1 at P_1. Jeep 1 goes to

B at a distance of 5/4. Jeep 2 goes from *A* to P'_1 at $x=1$. Jeep 1 makes a round trip from *B* to P'_1 to *B*, giving Jeep 2 enough fuel to reach *B*. Jeep 2 goes to *B*. Fuel required is 3.5 tanks.

(b) 4.12 units. Jeep 1 goes to P_1 at $x=0.02$ and stays there. Jeep 2 makes a round trip from *A* to P_1 to *A*, topping off Jeep 1 at P_1. Jeep 1 goes to P_2 at $x=0.34$ and stays there. Jeep 2 makes a round trip from *A* to P_2 to *A*, topping off Jeep 1 at P_2. Jeep 1 goes to B at a distance of 1.34. Jeep 2 goes from *A* to P'_2 at $x=1$. Jeep 1 makes a round trip from *B* to P'_2 to B, giving Jeep 2 enough fuel to reach P'_1 at $x=1.32$. Jeep 2 goes from P'_2 to P'_1. Jeep 1 makes a round trip from *B* to P'_1 to B, giving Jeep 2 enough fuel to reach *B*. Jeep 2 goes to *B*. Fuel required is 4.12 tanks.

(c) Distance is 1.39. Jeep 1 goes to P_1 at $x=0.17$ and stays there. Jeep 2 makes a round trip from *A* to P_1 to *A*, topping off Jeep 1 at P_1. Jeep 1 goes to P_2 at $x=0.39$ and stays there. Jeep 2 makes a round trip from *A* to P_2 to *A*, topping off Jeep 1 at P_2. Jeep 1 goes to *B* at a distance of 1.39. Jeep 2 goes from *A* to P'_2 at $x=1$. Jeep 1 makes a round trip from *B* to P'_2 to B, giving Jeep 2 enough fuel to reach P'_1 at $x=1.22$. Jeep 2 goes from P'_2 to P'_1. Jeep 1 makes a round trip from *B* to P'_1 to B, giving Jeep 2 enough fuel to reach *B*. Jeep 2 goes to *B*. Fuel required is 5.02 tanks.

(d) Distance is 263/180. Jeep 1 goes to P_1 at $x=0.15$ and stays there. Jeep 2 makes a round trip from *A* to P_1 to *A*, topping off Jeep 1 at P_1. Jeep 1 goes to P_2 at $x=23/60$ and stays there. Jeep 2 makes a round trip from *A* to P_2 to *A*, topping off Jeep 1 at P_2. Jeep 1 goes to P_3 at $x=83/180$ and stays there. Jeep 2 makes a round trip from *A* to P_3 to *A*, topping off Jeep 1 at P_3. Jeep 1 goes to *B* at a distance of 263/180. Jeep 2 goes from *A* to P'_3 at $x=1$. Jeep 1 makes a round trip from *B* to P'_3 to B, giving Jeep 2 enough fuel to reach P'_2 at $x=97/90$. Jeep 2 goes from P'_3 to P'_2. Jeep 1 makes a round trip from *B* to P'_2 to B, giving Jeep 2 enough fuel to reach P'_1 at $x=59/45$. Jeep 2 goes from P'_2 to P'_1. Jeep 1 makes a round trip from *B* to P'_1 to B, giving Jeep 2 enough fuel to reach *B*. Jeep 2 goes to *B*. Fuel required is 6.9 tanks.

(e) 7.65 units. Jeep 1 goes to P_1 at $x=0.275$ and stays there. Jeep 2 makes a round trip from *A* to P_1 to *A*, topping off Jeep 1 at P_1. Jeep 1 goes to P_2 at $x=0.425$ and stays there. Jeep 2 makes a round trip from *A* to P_2 to *A*, topping off Jeep 1 at P_2. Jeep 1 goes to P_3 at $x=0.475$ and stays there.

Jeep 2 makes a round trip from A to P_3 to A, topping off Jeep 1 at P_3. Jeep 1 goes to B at a distance of 1.475. Jeep 2 goes from A to P_3' at $x=1$. Jeep 1 makes a round trip from B to P_3' to B, giving Jeep 2 enough fuel to reach P_2' at $x=1.05$. Jeep 2 goes from P_3' to P_2'. Jeep 1 makes a round trip from B to P_2' to B, giving Jeep 2 enough fuel to reach P_1' at $x=1.2$. Jeep 2 goes from P_2' to P_1'. Jeep 1 makes a round trip from B to P_1' to B, giving Jeep 2 enough fuel to reach B. Jeep 2 goes to B. Fuel required is 7.65 tanks.

(f) Distance is 401/270. Jeep 1 goes to P_1 at $x=0.1$ and stays there. Jeep 2 makes a round trip from A to P_1 to A, topping off Jeep 1 at P_1. Jeep 1 goes to P_2 at $x=11/30$ and stays there. Jeep 2 makes a round trip from A to P_2 to A, topping off Jeep 1 at P_2. Jeep 1 goes to P_3 at $x=41/90$ and stays there. Jeep 2 makes a round trip from A to P_3 to A, topping off Jeep 1 at P_3. Jeep 1 goes to P_4 at $x=131/270$ and stays there. Jeep 2 makes a round trip from A to P_4 to A, topping off Jeep 1 at P_4. Jeep 1 goes to B at a distance of 401/270. Jeep 2 goes from A to P_4' at $x=1$. Jeep 1 makes a round trip from B to P_4' to B, giving Jeep 2 enough fuel to reach P_3' at $x=139/135$. Jeep 2 goes from P_4' to P_3'. Jeep 1 makes a round trip from B to P_3' to B, giving Jeep 2 enough fuel to reach P_2' at $x=151/135$. Jeep 2 goes from P_3' to P_2'. Jeep 1 makes a round trip from B to P_2' to B, giving Jeep 2 enough fuel to reach P_1' at $x=187/135$. Jeep 2 goes from P_2' to P_1'. Jeep 1 makes a round trip from B to P_1' to B, giving Jeep 2 enough fuel to reach B. Jeep 2 goes to B. Fuel required is 8.6 tanks.

(g) Distance is 3,361/2,250. Jeep 1 goes to P_1 at $x=0.332$ and stays there. Jeep 2 makes a round trip from A to P_1 to A, topping off Jeep 1 at P_1. Jeep 1 goes to P_2 at $x=0.444$ and stays there. Jeep 2 makes a round trip from A to P_2 to A, topping off Jeep 1 at P_2. Jeep 1 goes to P_3 at $x=361/750$ and stays there. Jeep 2 makes a round trip from A to P_3 to A, topping off Jeep 1 at P_3. Jeep 1 goes to P_4 at $x=1,111/2,250$ and stays there. Jeep 2 makes a round trip from A to P_4 to A, topping off Jeep 1 at P_4. Jeep 1 goes to B at a distance of 3,361/2,250. Jeep 2 goes from A to P_4' at $x=1$. Jeep 1 makes a round trip from B to P_4' to B, giving Jeep 2 enough fuel to reach P_3' at $x=1,139/1,125$. Jeep 2 goes from P_4' to P_3'. Jeep 1 makes a round trip from B to P_3' to B, giving Jeep 2 enough fuel to reach P'_2 at $x=1{,}181/1{,}125$. Jeep 2 goes from P'_3 to P'_2. Jeep 1 makes a round trip from B to P'_2 to B, giving Jeep 2 enough fuel to reach P'_1 at $x=1{,}307/1{,}125$. Jeep 2 goes from P'_2 to P'_1. Jeep 1 makes a round trip from B to P'_1 to B, giving Jeep 2 enough fuel to reach B. Jeep 2 goes to B. Fuel required is 9.992 tanks.

76
A Single One-Way Trip with Three Jeeps

(a) 4 units. Jeeps 1, 2 and 3 go to P_1 at $x=0.25$. Jeeps 1 and 2 top off and go from P_1 to P_2 at $x=0.5$. Jeep 1 goes from P_2 to B, at a distance of 1, and back to P_2. Jeeps 1 and 2 go from P_2 to P_1. Jeep 1 makes a round trip from P_1 to A to P_1. All jeeps go from P_1 to A. Fuel required is 4 tanks.

(b) 4.06 units. Jeeps 1, 2 and 3 go to P_1 at $x=0.1325$. Jeeps 1 and 2 top off and go from P_1 to P_2 at $x=0.255$. Jeep 1 goes from P_2 to P_3, at $x=0.505$. Jeep 3 goes from P_1 to A to P_2. Jeep 2 tops off and goes from P_2 to P_3. Jeep 1 tops off and makes a round trip from P_3 to B at $x=1.005$ to P_3. Jeeps 1 and 2 go from P_3 to P_1, taking on fuel from Jeep 3 at P_2. Jeep 3 goes from P_2 to A to P_1. All jeeps go from P_1 to A. Fuel required is 4.06 tanks.

(c) 4.12 units. Jeeps 1, 2 and 3 go to P_1 at $x=0.14$. Jeeps 1 and 2 top off and go from P_1 to P_2 at $x=0.26$. Jeep 1 goes from P_2 to P_3, at $x=0.51$. Jeep 3 goes from P_1 to A to P_2. Jeep 2 tops off and goes from P_2 to P_3. Jeep 1 tops off and makes a round trip from P_3 to B at $x=1.01$ to P_3. Jeeps 1 and 2 go from P_3 to P_1. Jeep 3 goes from P_2 to A to P_1, taking on fuel from Jeep 3 at P_2. All jeeps go from P_1 to A. Fuel required is 4.12 tanks.

(d) 4.48 units. Jeeps 1, 2 and 3 go to P_1 at $x=0.185$. Jeeps 1 and 2 top off and go from P_1 to P_2 at $x=0.29$. Jeep 1 goes from P_2 to P_3, at $x=0.54$. Jeep 3 goes from P_1 to A to P_2. Jeep 2 tops off and goes from P_2 to P_3. Jeep 1 tops off and makes a round trip from P_3 to B at $x=1.04$ to P_3. Jeeps 1 and 2 go from P_3 to P_1, taking on fuel from Jeep 3 at P_2. Jeep 3 goes from P_2 to A to P_1. All jeeps go from P_1 to A. Fuel required is 4.48 tanks.

(e) Distance is 16/15. Jeeps 1, 2 and 3 go to P_1 at $x=9/40$. Jeeps 1 and 2 top off and go from P_1 to P_2 at $x=9/40+11/120=19/60$. Jeep 1 goes from P_2 to P_3, at $x=19/60+1/4=17/30$. Jeep 3 goes from P_1 to A to P_2. Jeep 2 tops off and goes from P_2 to P_3. Jeep 1 tops off and makes a round trip from P_3 to B at $x=16/15$ to P_3. Jeeps 1 and 2 go from P_3 to P_1, taking on fuel from Jeep 3 at P_2. Jeep 3 goes from P_2 to A to P_1. All jeeps go from P_1 to A. Fuel required is 4.8 tanks.

(f) Distance is 13/12. Jeeps 1, 2 and 3 go to P_1 at $x=0.25$. Jeeps 1 and 2 top off and go from P_1 to P_2 at $x=0.25+1/12=1/3$. Jeep 1 goes from P_2 to P_3, at $x=1/3+1/4=7/12$. Jeep 3 goes from P_1 to A to P_2. Jeep 2 tops off and goes from P_2 to P_3. Jeep 1 tops off and makes a round trip from P_3 to B at $x=13/12$ to P_3. Jeeps 1 and 2 go from P_3 to P_1, taking on fuel from Jeep 3 at P_2. Jeep 3 goes from P_2 to A to P_1. All jeeps go from P_1 to A. Fuel required is 5 tanks.

Chapter 10: MathDice Puzzles

77
Use 0, 1 and 2 in Ascending Order to Make

(a) $12=0+12$
(b) $13=0!+12$
(c) $24=(0!+1+2)!=[(0!+1)\times2]!$

78
Use 1, 2 and 3 in Ascending Order to Make

(a) $12=1\times2\times3!=(1+2)!+3!$

(b) $13=1+2\times3!$

(c) $15=12+3$

(d) $24=1+23=(12/3)!=(1^2+3)!$

(e) $27=(1+2)^3$

(f) $36=12\times3=(1+2)!\times3!$

(g) $64=1\times2^{3!}=(1\times2)^{3!}$

(h) $72=12\times3!$

79
Use 2, 3 and 4 in Ascending Order to Make

(a) $14 = 2 \times (3+4) = 2 + 3 \times 4$

(b) $16 = 2 \times 3! + 4 = (2/3) \times 4! = 2^{3!}/4$

(c) $20 = (2+3) \times 4 = 2 - 3! + 4!$

(d) $24 = 2 \times 3 \times 4 = (2 + 3! - 4)!$

(e) $30 = 2 \times 3 + 4! = (2+3)!/4 = (2 \times 3)!/4!$

(f) $36 = 2 \times 3! + 4! = 2 + 34$

(g) $40 = 2^{3!} - 4!$

(h) $47 = 23 + 4!$

(i) $54 = 2 \times (3 + 4!)$

(j) $60 = 2 \times 3!!/4! = 2 \times (3! + 4!)$

(k) $68 = 2 \times 34$

(l) $74 = 2 + 3 \times 4!$

(m) $83 = 2 + 3^4$

(n) $88 = 2^{3!} + 4!$

(o) $96 = (2+3)! - 4!$

80
Use 3, 4 and 5 in Ascending Order to Make

(a) $12 = 3 + 4 + 5$

(b) $15 = (3/4!) \times 5! = 3! + 4 + 5$

(c) $16 = 3!!/45$

(d) $17 = 3 \times 4 + 5$

(e) $19 = 3! \times 4 - 5$

(f) $22 = 3 + 4! - 5$

(g) $24 = (3 - 4 + 5)! = 3!! \times 4/5! = 3! \times 4! - 5!$

(h) $25 = 3! + 4! - 5 = 3!!/4! - 5$

(i) $26 = 3! + 4 \times 5$

(j) $27 = 3 \times (4 + 5)$

(k) $29 = 34 - 5 = 3! \times 4 + 5$

(l) $30 = (3!/4!) \times 5!$

(m) $32 = 3 + 4! + 5 = (3! - 4)^5$

(n) $35 = (3+4) \times 5 = 3! + 4! + 5 = 3!!/4! + 5$

(o) $36 = 3!!/4/5$

(p) $39 = 34 + 5$

(q) $42 = (3+4)!/5!$

(r) $50 = (3! + 4) \times 5$

(s) $51 = 3! + 45$

(t) $54 = 3! \times (4 + 5)$

(u) $57 = 3 \times (4! - 5)$

(v) $60 = 3 \times 4 \times 5 = 3!!/4 - 5!$

(w) $67 = 3 \times 4! - 5$

(x) $76 = 3^4 - 5$

(y) $77 = 3 \times 4! + 5$

(z) $80 = 3!!/(4 + 5)$

(a1) $86 = 3^4 + 5$

(b1) $87 = 3 \times (4! + 5)$

(c1) $90 = (3/4) \times 5!$

(d1) $99 = 3 - 4! + 5!$

81
Use 4, 5 and 6 in Ascending Order to Make

(a) $13 = 4! - 5 - 6$

(b) $20 = 4! \times 5/6$

(c) $24 = (4! + 5!)/6 = 4 + 5!/6 = 4/(5!/6!)$

(d) $25 = 4! - 5 + 6$

(e) $26 = 4 \times 5 + 6$

(f) $34 = 4 + 5 \times 6$

(g) $35 = 4! + 5 + 6$

(h) $39 = 45 - 6$

(i) $44 = 4 \times (5 + 6) = 4! + 5!/6$

(j) $51 = 45 + 6$

(k) $54 = (4 + 5) \times 6 = 4! + 5 \times 6$

(l) $60 = 4 + 56$

(m) $80 = 4 \times 5!/6 = 4! + 56$

82
Use 5, 6 and 7 in Ascending Order to Make

(a) $18 = 5 + 6 + 7 = (5! + 6)/7$

(b) $23 = 5 \times 6 - 7$

(c) $24 = (5 + 6 - 7)!$

(d) $27 = 5!/6 + 7$

(e) $35 = (5/6!) \times 7!$

(f) $37 = 5 \times 6 + 7$

(g) $47 = 5 + 6 \times 7$

(h) $49 = 56 - 7$

(i) $53 = 5! - 67$

(j) $63 = 56 + 7$

(k) $65 = 5 \times (6 + 7)$

(l) $72 = 5 + 67$

(m) $77 = (5 + 6) \times 7$

(n) $78 = 5! - 6 \times 7$

83
Use 6, 7 and 8 in Ascending Order to Make

(a) $21 = 6 + 7 + 8$

(b) $34 = 6 \times 7 - 8$

(c) $48 = (6/7!) \times 8! = 6!/(7 + 8)$

(d) $50 = 6 \times 7 + 8$

(e) $59 = 67 - 8$

(f) $62 = 6 + 7 \times 8$

(g) $75 = 67 + 8$

(h) $90 = 6! \times 7!/8! = 6 \times (7 + 8) = 6! - 7!/8$

84
Use 7, 8 and 9 in Ascending Order to Make

(a) $24 = 7+8+9$

(b) $47 = 7 \times 8 - 9$

(c) $63 = (7/8!) \times 9!$

(d) $65 = 7 \times 8 + 9$

(e) $69 = 78 - 9$

(f) $70 = 7!/8/9$

(g) $79 = 7 + 8 \times 9$

(h) $87 = 78 + 9$

(i) $96 = 7 + 89$

85
Use 0, 2 and 4 in Ascending Order to Make

(a) $17 = 0 + 2^4$

(b) $23 = 0! - 2 + 4!$

(c) $25 = 0!^2 + 4! = 0! + 24$

(d) $30 = (0!+2)! + 4! = (0!+2)!!/4!$

(e) $49 = 0! + 2 \times 4!$

(f) $72 = (0!+2) \times 4!$

(g) $81 = (0!+2)^4$

86
Use 1, 3 and 5 in Ascending Order to Make

(a) $12 = 1 + 3! + 5$

(b) $15 = 1 \times 3 \times 5$

(c) $16 = 1 + 3 \times 5$

(d) $19 = (1+3)! - 5$

(e) $20 = (1+3) \times 5 = (1/3!) \times 5!$

(f) $29 = (1+3)! + 5$

(g) $30 = 1 \times 3! \times 5$

(h) $31 = 1 + 3! \times 5$

(i) $40 = (1/3) \times 5!$

(j) $42 = (1+3!)!/5!$

(k) $65 = 13 \times 5$

87
Use 2, 4 and 6 in Ascending Order to Make

(a) $16 = 2^{4!/6}$

(b) $20 = 2 + 4! - 6 = 2 \times (4 + 6)$

(c) $22 = 2^4 + 6$

(d) $24 = (24/6)!$

(e) $26 = 2 + 4 \times 6 = 2 + (4!/6)!$

(f) $30 = 24 + 6$

(g) $32 = 2 + 4! + 6$

(h) $36 = (2 + 4) \times 6 = 2 \times (4! - 6)$

(i) $42 = 2 \times 4! - 6$

(j) $48 = 2 \times 4 \times 6 = 2 \times (4!/6)! = 2 + 46$

(k) $54 = 2 \times 4! + 6$

(l) $56 = (2 \times 4)!/6!$

(m) $60 = (2/4!) \times 6! = 2 \times (4! + 6)$

(n) $64 = (2 - 4)^6$

(o) $96 = 2^4 \times 6$

88
Use 3, 5 And 7 in Ascending Order to Make

(a) $13 = 3!!/5! + 7$

(b) $15 = 3 + 5 + 7$

(c) $18 = (3! + 5!)/7 = 3! + 5 + 7$

(d) $22 = 3 \times 5 + 7$

(e) $23 = 3! \times 5 - 7$

(f) $24 = (3! + 5 - 7)!$

(g) $28 = 35 - 7$

(h) $36 = 3 \times (5 + 7)$

(i) $37 = 3! \times 5 + 7$

(j) $38 = 3 + 5 \times 7$

(k) $41 = 3! + 5 \times 7$

(l) $42 = 35 + 7 = (3!!/5!) \times 7$

(m) $56 = (3 + 5) \times 7$

(n) $60 = 3 + 57 = 3!!/(5 + 7)$

(o) $63 = 3! + 57$

(p) $72 = 3! \times (5 + 7)$

(q) $77 = (3! + 5) \times 7$

89
Use 4, 6 and 8 in Ascending Order to Make

(a) $12 = (4!/6) + 8$

(b) $16 = 4 \times 6 - 8 = (4!/6)! - 8$

(c) $18 = 4 + 6 + 8 = 4! \times 6/8$

(d) $22 = 4! + 6 - 8$

(e) $26 = 4! - 6 + 8$

(f) $32 = 4 \times 6 + 8 = (4!/6) \times 8 = (4!/6)! + 8$

(g) $38 = 46 - 8 = 4! + 6 + 8$

(h) $52 = 4 + 6 \times 8$

(i) $54 = 46 + 8$

(j) $56 = 4 \times (6 + 8)$

(k) $72 = 4! + 6 \times 8$

(l) $80 = (4 + 6) \times 8$

(m) $90 = (4 + 6)!/8!$

(n) $92 = 4! + 68$

(o) $93 = (4! + 6!)/8$

90
Use 5, 7 and 9 in Ascending Order to Make

(a) $21 = 5 + 7 + 9$

(b) $26 = 5 \times 7 - 9$

(c) $41 = 5! - 79$

(d) $44 = 5 \times 7 + 9$

(e) $48 = 57 - 9$

(f) $57 = 5! - 7 \times 9$

(g) $66 = 57 + 9$

(h) $68 = 5 + 7 \times 9$

(i) $80 = 5 \times (7 + 9)$

(j) $84 = 5 + 79$

91
Use 0, 3 and 6 in Ascending Order to Make

(a) $13 = 0! + 3! + 6$

(b) $19 = 0! + 3 \times 6$

(c) $24 = (0! + 3) \times 6 = [(0! + 3)!/6]!$

(d) $30 = (0! + 3)! + 6$

(e) $37 = 0! + 36 = 0! + 3! \times 6$

(f) $42 = (0! + 3!) \times 6$

(g) $64 = (0! - 3)^6$

92
Use 1, 4 and 7 in Ascending Order to Make

(a) $11 = 1 \times (4 + 7) = 1 \times 4 + 7$

(b) $12 = 1 + 4 + 7$

(c) $18 = 1 + 4! - 7$

(d) $21 = 14 + 7$

(e) $24 = (1 - 4 + 7)!$

(f) $31 = 1 \times 4! + 7 = (1 \times 4)! + 7$

(g) $32 = 1 + 4! + 7$

(h) $35 = (1 + 4) \times 7$

(i) $47 = 1 \times 47$

(j) $48 = 1 + 47$

(k) $98 = 14 \times 7$

93
Use 2, 5 and 8 in Ascending Order to Make

(a) $15 = 2 + 5 + 8$

(b) $17 = 25 - 8 = 2 + 5!/8$

(c) $18 = 2 \times 5 + 8$

(d) $24 = 2^5 - 8$

(e) $26 = 2 \times (5 + 8)$

(f) $30 = 2 \times 5!/8$

(g) $33 = 25 + 8$

(h) $40 = 2^5 + 8$

(i) $42 = 2 + 5 \times 8$

(j) $56 = (2 + 5) \times 8$

(k) $60 = 2 + 58$

(l) $80 = 2 \times 5 \times 8$

(m) $90 = (2 \times 5)!/8!$

94
Use 3, 6 and 9 in Ascending Order to Make

(a) $18=3+6+9$

(b) $21=3!+6+9$

(c) $24=(36/9)!=[(3!\times6)/9]!$

(d) $27=3\times6+9=36-9$

(e) $45=36+9=3\times(6+9)=3!\times6+9$

(f) $48=3!!/(6+9)$

(g) $75=3!+69$

(h) $81=(3+6)\times9=3^6/9$

(i) $83=3+6!/9$

(j) $86=3!+6!/9$

(k) $90=3!\times(6+9)$

95
Use 0, 4 and 8 in Ascending Order to Make

(a) $13=0!+4+8$

(b) $15=(0!+4)!/8$

(c) $17=0!+4!-8$

(d) $33=0!+4!+8$

(e) $40=(0!+4)\times8$

(f) $48=0!\times48=0+48$

(g) $49=0!+48$

96
Use 1, 5 and 9 in Ascending Order to Make

(a) $15=1+5+9$

(b) $24=15+9$

(c) $46=1+5\times9$

(d) $54=(1+5)\times9$

(e) $59=1\times59$

(f) $60=1+59$

(g) $80=(1+5)!/9$

97
Use 1, 2 and 3 in Descending Order to Make

(a) $11=3!\times2-1$

(b) $18=3\times(2+1)!$

(c) $23=(3!-2)!-1$

(d) $27=3^{2+1}=3!+21$

(e) $32=32\times1=32/1=32^1$

(f) $35=3!^2-1$

(g) $37=3!^2+1$

(h) $63=3\times21$

98
Use 1, 2 and 4 in Descending Order to Make

(a) $11 = 4!/2 - 1$

(b) $15 = 4^2 - 1$

(c) $18 = 4! - (2 + 1)!$

(d) $21 = 4! - 2 - 1$

(e) $25 = 4! + 2 - 1 = 4 + 21$

(f) $26 = 4! + 2 \times 1 = 4! + 2/1 = 4! + 2^1$

(g) $30 = 4! + (2 + 1)!$

(h) $45 = 4! + 21$

(i) $47 = 4! \times 2 - 1$

(j) $64 = 4^{2+1}$

99
Use 1, 2 and 5 in Descending Order to Make

(a) $11 = 5 + (2 + 1)! = 5 \times 2 + 1$

(b) $20 = 5!/(2 + 1)!$

(c) $24 = (5 - 2 + 1)! = 5^2 - 1$

(d) $26 = 5 + 21 = 5^2 + 1$

(e) $30 = 5 \times (2 + 1)!$

(f) $40 = 5!/(2 + 1)$

(g) $51 = 52 - 1$

(h) $61 = 5!/2 + 1$

(i) $99 = 5! - 21$

100
Use 3, 5 and 6 in Descending Order to Make

(a) $12 = 6 \times (5 - 3) = 6!/5! + 3!$

(b) $18 = (6!/5!) \times 3$

(c) $21 = 6 + 5 \times 3 = (6 + 5!)/3!$

(d) $24 = (6 - 5 + 3)! = 6 \times 5 - 3! = 6!/5/3!$

(e) $26 = 6 + 5!/3!$

(f) $33 = 6 \times 5 + 3 = (6 + 5) \times 3$

(g) $36 = 65 - 3 = (6!/5!) \times 3! = 6 \times 5 + 3!$

(h) $46 = 6 + 5!/3$

(i) $48 = 6 \times (5 + 3) = 6!/5/3$

(j) $59 = 6 + 53 = 65 - 3!$

(k) $66 = 6 \times (5 + 3!)$

(l) $90 = 6 \times 5 \times 3 = 6!/(5 + 3)$

(m) $100 = (6! - 5!)/3!$

101
Use 3, 6 and 7 in Descending Order to Make

(a) $13 = 7 + (6-3)! = 7!/6! + 3!$

(b) $21 = 7 \times (6-3) = (7!/6!) \times 3$

(c) $24 = (7-6+3)! = (7!/6!-3)!$

(d) $39 = (7+6) \times 3 = 7 \times 6 - 3$

(e) $42 = 7 \times (6-3)! = (7!/6!) \times 3!$

(f) $49 = 7^{6/3}$

(g) $70 = 76 - 3! = 7 + 63$

(h) $78 = (7+6) \times 3!$

(i) $80 = 7!/63$

(j) $82 = 76 + 3!$

(k) $84 = 7 \times (6+3!)$

102
Use 3, 6 and 8 in Descending Order to Make

(a) $14 = 8 + (6-3)!$

(b) $24 = [(8/6) \times 3]! = 8 \times (6-3)$

(c) $28 = 8!/(6!+3!!)$

(d) $48 = 8 \times (6-3)!$

(e) $50 = 8!/6! - 3!$

(f) $55 = (8!-6!)/3!!$

(g) $56 = 8!/(6-3)!!$

(h) $57 = (8!+6!)/3!!$

(i) $64 = 8^{6/3} = (8-6)^{3!}$

(j) $80 = 86 - 3!$

(k) $96 = 8 \times (6+3!)$

103
Use 3, 4 and 7 in Descending Order to Make

(a) $11 = 7 + 4!/3!$

(b) $18 = (7-4)! \times 3 = (7-4) \times 3!$

(c) $25 = 7 + 4! - 3! = 7 \times 4 - 3$

(d) $27 = (7-4)^3$

(e) $28 = 7 + 4! - 3 = 7! \times 4/3!! = 7 \times 4!/3!$

(f) $31 = 7 \times 4 + 3 = 7 + (4!/3!)! = 7 + 4 \times 3!$

(g) $35 = 7!/4!/3!$

(h) $36 = (7-4)! \times 3!$

(i) $37 = 7 + 4! + 3!$

(j) $70 = 7 \times (4+3!)$

(k) $71 = 74 - 3 = 7 + 4^3$

104
Use 3, 4 and 5 in Descending Order to Make

(a) $11 = 5!/4! + 3!$

(b) $12 = 5 + 4 + 3 = 5!/(4 + 3!)$

(c) $15 = 5 + 4 + 3! = (5!/4!) \times 3$

(d) $16 = (5! - 4!)/3!$

(e) $20 = 5 \times 4!/3! = (5!/4!)!/3!$

(f) $29 = 5 + (4!/3!)! = 5 + 4 \times 3!$

(g) $30 = (5!/4!) \times 3!$

(h) $32 = 5 + 4! + 3 = (5! - 4!)/3$

(i) $40 = 5 \times 4!/3 = (5!/4!)!/3$

(j) $51 = 54 - 3$

(k) $56 = 5! - 4^3$

(l) $90 = (5!/4) \times 3 = 5! - 4! - 3! = 5 \times (4! - 3!)$

(m) $96 = 5! - 4 \times 3! = 5! - (4!/3!)!$

105
Use 3, 5 and 8 in Descending Order to Make

(a) $12 = (8 - 5)! + 3!$

(b) $18 = (8 - 5)! \times 3 = (8 - 5) \times 3!$

(c) $24 = [8/(5 - 3)]!$

(d) $27 = (8 - 5)^3$

(e) $28 = 8 + 5!/3!$

(f) $32 = 8^{5/3}$

(g) $48 = 8/(5!/3!!) = 8 + 5!/3$

(h) $56 = 8!/5!/3!$

(i) $64 = 8 \times (5 + 3) = 8^{5-3}$

(j) $78 = (8 + 5) \times 3!$

(k) $88 = 85 + 3 = 8 \times (5 + 3!)$

106
Use 3, 7 and 8 in Descending Order to Make

(a) $14 = 8!/7! + 3!$

(b) $15 = 8 + 7!/3!!$

(c) $24 = (8 - 7 + 3)! = (8!/7!) \times 3$

(d) $29 = 87/3 = 8 + 7 \times 3$

(e) $49 = (8! - 7!)/3!!$

(f) $56 = 8 \times 7!/3!!$

(g) $63 = (8! + 7!)/3!!$

(h) $81 = 87 - 3! = 8 + 73$

(i) $90 = 87 + 3 = (8 + 7) \times 3!$

107
Use 3, 4 and 8 in Descending Order to Make

(a) $12 = (8-4) \times 3 = (8/4) \times 3! = 8 + 4!/3!$

(b) $16 = 8^{4/3} = 8 + 4!/3$

(c) $24 = (8-4!/3!)! = [(8+4)/3]! = (8-4) \times 3! = [(8-4)!/3!]!$

(d) $28 = 84/3$

(e) $30 = (8-4)! + 3!$

(f) $32 = 8 \times 4!/3! = 8 + 4 \times 3! = 8 + (4!/3!)!$

(g) $35 = 8 \times 4 + 3 = 8 + 4! + 3$

(h) $64 = (8-4)^3 = (8/4)^{3!} = 8 \times 4!/3$

(i) $72 = (8-4)! \times 3 = (8+4) \times 3! = 8 + 4^3$

(j) $80 = 8 \times (4+3!) = 8 + 4! \times 3$

(k) $96 = 8 \times 4 \times 3 = (8+4!) \times 3$

108
Use 3, 4 and 6 in Descending Order to Make

(a) $18 = 6 \times 4 - 3!$

(b) $27 = 6 \times 4 + 3 = 6 + 4! - 3 = 6!/4! - 3$

(c) $33 = 6 + 4! + 3 = 6!/4! + 3$

(d) $36 = 6 + 4! + 3! = 6!/4! + 3!$

(e) $40 = 6!/(4! - 3!)$

(f) $58 = 64 - 3!$

(g) $60 = (6+4) \times 3! = 6!/4/3$

(h) $64 = (6-4)^{3!}$

(i) $70 = 64 + 3! = 6 + 4^3$

(j) $72 = 6 \times 4 \times 3 = 6!/(4+3!)$

(k) $90 = (6+4!) \times 3 = (6!/4!) \times 3$

109
Use 4, 7 and 9 in Descending Order to Make

(a) $15 = 9 + (7-4)!$

(b) $16 = (9-7)^4$

(c) $18 = 9!/7!/4$

(d) $24 = [(9+7)/4]!$

(e) $26 = 9 - 7 + 4!$

(f) $39 = 9 \times 7 - 4!$

(g) $48 = (9-7) \times 4! = 9!/7! - 4!$

(h) $68 = 9!/7! - 4$

(i) $73 = 97 - 4!$

(j) $83 = 9 + 74$

(k) $99 = 9 \times (7+4)$

110
Use 4, 8 and 9 in Descending Order to Make

(a) $13 = 9 + 8 - 4 = 9!/8! + 4$

(b) $24 = (9 - 8) \times 4! = [(9 - 8) \times 4]!$

(c) $27 = (9/8) \times 4!$

(d) $33 = 9!/8! + 4! = 9 + (8 - 4)!$

(e) $36 = 9 \times (8 - 4) = (9!/8!) \times 4$

(f) $41 = 9 + 8 + 4! = 9 + 8 \times 4$

(g) $68 = (9 + 8) \times 4 = 9 \times 8 - 4$

(h) $74 = 98 - 4!$

(i) $81 = 9^{8/4}$

(j) $93 = 9 + 84$

(k) $96 = 9 \times 8 + 4!$

111
Use 3, 4 and 9 in Descending Order to Make

(a) $11 = 9 - 4 + 3! = (9 + 4!)/3$

(b) $17 = 9 + 4!/3$

(c) $20 = (9 - 4)!/3!$

(d) $21 = 9!/4!/3! = 9 + 4 \times 3$

(e) $27 = 9 + 4! - 3!$

(f) $33 = 9 + (4!/3!)! = 9 \times 4 - 3 = 9 + 4 \times 3!$

(g) $39 = (9 + 4) \times 3 = 9 + 4! + 3! = 9 \times 4 + 3$

(h) $40 = (9 - 4)!/3$

(i) $72 = 9 \times 4!/3 = 9!/(4 + 3)!$

(j) $73 = 9 + 4^3$

(k) $88 = 94 - 3!$

(l) $99 = (9 + 4!) \times 3$

(m) $100 = 94 + 3!$

112
Stimulating Singles

(a) $15 = 2 \times 6/.8$

(b) $25 = 9 \times .6^{-2}$

(c) $30 = 27/.9$

(d) $7 = 8 - .8 - .2$

(e) $11 = (9 - .2)/.8$

(f) $25 = 3/(.3 \times .4)$

(g) $13 = 6/.3 - 7$

(h) $17 = (6 - .9)/.3$

(i) $9 = 8 + .7 + .3$

(j) $24 = (8 - .8)/.3$

(k) $9 = (4 - .4)/.4$

(l) $16 = 4/(.5 \times .5)$

(m) $28 = .5^{-5} - 4$

(n) $30 = 4 \times (8 - .5)$

(o) $23 = 9/.4 + .5$

(p) $64 = 4 \wedge (9^{.5})$

(q) $11 = 6 + 4/.8$

(r) $18 = 9/(.9 - .4)$

(s) $16 = .5 \times .5^{-5}$

(t) $11 = (6 - .5)/.5$

(u) $26 = .5^{-5} - 6$

(v) $11 = 8 + 9^{.5}$

(w) $64 = .5^{-9}/8$

113
Perplexing Pairs

(a) $7=1/.2+2=2+.2^{-1}$

(b) $18=2/.1-2=(2-.2)/.1$

(c) $8=3\times3-1=(3-1)^3$

(d) $6=4+3-1=3/(.1+.4)$

(e) $10=8+1/.5=8+.5^{-1}$

(f) $16=1+6+9=1+9/.6$

(g) $32=29+3=2+9/.3$

(h) $14=4+2\times5=.5^{-4}-2$

(i) $22=4+2\times9=(9-.2)/.4$

(j) $64=2^5/.5=2\times.5^{-5}$

(k) $25=.2^{6-8}=(.8-.6)^{-2}$

(l) $7=2\times8-9=8\times.9-.2$

(m) $31=3^3+4=34-3$

(n) $24=3\times(3+5)=3\times.5^{-3}$

(o) $30=3\times9+3=3^3/.9$

(p) $4=3-4+5=.5^{-3}-4$

(q) $8=3+9-4=(.9-.4)^{-3}$

(r) $14=3/.5+8=(5-.8)/.3$

(s) $18=6\times(6-3)=(6-.6)/.3$

(t) $8=4\times4\times.5=4\times4^5$

(u) $12=4/.5+4=.5^{-4}-4$

(v) $18=(4+5)/.5=4\times(5-.5)$

(w) $16=.5^{-6}/4=4+6/.5$

(x) $32=(8/4)^5=.5^{-4/.8}$

(y) $25=5\times(9-4)=.5^{-4}+9$

114
Troublesome Trios

(a) $8=1/.1-2=(1-.2)/.1=.1^{-1}-2$

(b) $13=8+1/.2=21-8=8+.2^{-1}$

(c) $8=1+3+4=4\times(3-1)=(.1+.4)^{-3}$

(d) $15=6/(2\times.2)=6/(.2+.2)=.6\times.2^{-2}$

(e) $2=8-2\times3=(8-2)/3=3-.2-.8$

(f) $10=3+9-2=3^2/.9=.9\times.3^{-2}$

(g) $10=2/.5+6=6/.5-2=.5^{-2}+6$

(h) $32=5\times6+2=.5\times2^6=.5^{-6}/2$

(i) $28=2\times7/.5=7\times.5^{-2}=7/.25$

(j) $9=3+3+3=3^3/3=(3-.3)/.3$

(k) $16=(6/3)^4=4^{6/3}=6/.3-4$

(l) $13=3+5+5=5/.5+3=5+.5^{-3}$

(m) $4=(5+7)/3=5-.3-.7=3/.75$

(n) $15=3+5+7=3/(.7-.5)=7+.5^{-3}$

(o) $32=8^{5/3}=.5^{3-8}=(.8-.3)^{-5}$

(p) $10=5\times(6-4)=6/.4-5=.5^{-4}-6$

(q) $8=5+7-4=4\times(7-5)=.5^{4-7}$

(r) $9=7+.4\times5=7+4^{.5}=.5^{-4}-7$

(s) $20=5\times(8-4)=8/.5+4=.5^{-4}/.8$

(t) $25=5^{7-5}=5/(.7-.5)=.5^{-5}-7$

(u) $4=6/.5-8=(8-6)/.5=.5^{6-8}$

(v) $25=5^{9-7}=5/(.9-.7)=9/.5+7$

115
Menacing Multiples

(a) $5=2\times6-7=7/(2-.6)=(7-6)/.2=.2^{6-7}$

(b) $27=3\times(3+6)=36^{-3}=(6-3)^3=33-6$

(c) $32=4\times(3+5)=.5\times4^3=4\times.5^{-3}=4^{3-.5}$

(d) $6=3\times5-9=9/(3\times.5)=3+9^{.5}=.5\times(3+9)$

(e) $2=8/5+.4=.5\times(8-4)=(8-4)^{.5}=.5^{-4}/8$

(f) $8=6+7-5=7/.5-6=6/.75=56/7$

(g) $6=5+8-7=.8\times(7+.5)=7/.5-8=(5-.8)/.7$

(h) $32=5\times8-8=5\times8\times.8=(8+8)/.5=.5^{-8}/8$

(i) $64=4\times16=1\times64=64/1=64^1=6/.1+4=(.1+.4)^{-6}$

(j) $2=8-4-2=(8-4)/2=2^4/8=4^2/8=(2-.4)/.8=.2\times8+.4=(4\times8)^2$

(k) $16=.5\times2^5=.5^{-2/.5}=(.5\times.5)^{-2}=.5^{-5}/2=.5^\wedge-(.5^{-2})$

(l) $32=5^2+7=25+7=5/.2+7=.5^{2-7}=(.7-.2)^{-5}$

(m) $5=9-2/.5=2/(.9-.5)=9/2+.5=2+9^{.5}=9-.5^{-2}$

(n) $16=4+4+8=4\times(8-4)=4^{8/4}=(8/4)^4=8/.4-4$

116
Largest Possible

(a) Let $a=.1^{-4}=10{,}000$ and $b=.2^{-a}=5^{1000.}$ Then $E_1=.3^{-b}$ is largest. Ln(ln(E_1)) $\approx$ 16,095. Ln(x) is the natural logarithm of x.

(b) Let $p=^{-\sqrt[1]{.2}}=5^{10}=9{,}765{,}625$ and let $q=4^p$. Then $E_2=.3^{-q}$ is largest. Ln(ln(E_2)) $\approx$ 13,538,031.

Acknowledgments

1	Word Mystery	Laurie Brokenshire
2	Salary Secrecy	Bob Wainwright
3	Relations Puzzles	Dick Hess
4	Slider	Neil Bickford
5	Fastest Serve	Dick Hess
6	Population Explosion	Dick Hess
7	Catenary	Joop van der Vaart
6	Mining on Rigel IV	Dick Hess
9	Linking Points	Dick Hess
10	Right Triangles	Dick Hess
11	The Clipped Polyhedron	G. Galperin
12	The Papered Boxes	Dick Hess
13	The Almost Rectangular Lake	Leon Bankoff
14	The Trifurcated Diamond	Dick Hess
15	Square Dissection	Dick Hess
16	Pandigital Purchases	Dick Hess
17	Sudoku Variant	Bob Wainwright
18	Pandigital Sums	Dick Hess
19	Even and Odd	World contest problems
20	Integer Oddity	Dick Hess
21	A 10-Digit Number	Dick Hess
22	Cake Division	Dick Hess
23	Prisoner's Escape	Andy Liu/Dick Hess
24	Logical Question	Dick Hess

Printed in the United States
By Bookmasters